Advances in

Friedel-Crafts Acylation Reactions

Catalytic and Green Processes

Advances in

Friedel-Crafts Acylation Reactions

Catalytic and Green Processes

Giovanni Sartori *and* Raimondo Maggi

CRC Press
Taylor & Francis Group
Boca Raton London New York

CRC Press is an imprint of the
Taylor & Francis Group, an **informa** business

CRC Press
Taylor & Francis Group
6000 Broken Sound Parkway NW, Suite 300
Boca Raton, FL 33487-2742

First issued in paperback 2017

© 2010 by Taylor and Francis Group, LLC
CRC Press is an imprint of Taylor & Francis Group, an Informa business

No claim to original U.S. Government works

ISBN 13: 978-1-138-11384-8 (pbk)
ISBN 13: 978-1-4200-6792-7 (hbk)

Library of Congress Cataloging-in-Publication Data

Sartori, Giovanni, 1944-
 Advances in Friedel-Crafts acylation reactions : catalytic and green processes / Giovanni Sartori, Raimondo Maggi.
  p. cm.
 Includes bibliographical references and index.
 ISBN 978-1-4200-6792-7 (hardcover : alk. paper)
 1. Friedel-Crafts reaction. I. Maggi, Raimondo. II. Title.

QD281.A5S27 2010
547'.2--dc22

2009034801

Visit the Taylor & Francis Web site at
http://www.taylorandfrancis.com

and the CRC Press Web site at
http://www.crcpress.com

Dedication

In memory of Prof. Giuseppe (Peppino) Casnati

Contents

Preface

The fundamental treatise entitled *Friedel–Crafts and Related Reactions* was edited and, in part, written in five volumes by George A. Olah from 1963 to 1965. A single volume was further published by the same author in 1972, discussing advances achieved in the field in the period 1965–1972, including mechanistic aspects of the Friedel–Crafts reactions.

Since 1972, updates on Friedel–Crafts alkylation, aromatic aldehyde synthesis, nitration, etc., were regularly published in books and extensive reviews. However, improvements in Friedel–Crafts acylations were scantly considered, despite the great practical application of the aromatic ketones in different fields of the fine and pharmaceutical chemistry.

In 2006 we published a review article dealing with the "Use of Solid Catalysts in Friedel–Crafts Acylation Reactions" (*Chem. Rev.*, 2006, *106*, 1077–1104). It was thus with great enthusiasm that we accepted the publisher's invitation to produce a book on the advances in Friedel–Crafts acylation, focusing on truly catalytic and green processes.

However, in researching the literature, we had to face the great number and varied typology of scientific articles available, ranging from merely synthetic notes to extremely detailed studies, frequently facing engineering problems and sometimes getting conflicting conclusions.

Moreover, because of the great interest in catalytic Friedel–Crafts acylations, studies on this topic were continuously progressing, and consequently, it was very difficult to recognize the best results.

Nevertheless, the even difficult comparative analysis of articles and patents allowed some exciting conclusions concerning, in particular, the feasibility of very efficient catalytic Friedel–Crafts acylation reactions.

Now we are convinced that a monograph on the advances in Friedel–Crafts acylation reactions is timely, especially in view of the present-day requirements in green synthetic processes.

This work incorporates all acid-catalyzed Friedel–Crafts-like acylation reactions. Thus, classic Lewis and Brönsted acid types are considered together with more innovative and advanced multicomponent super-acid catalysts, ranging from rare earth triflates or triflimides and their combination with ionic liquids to metal-promoted zeolites and zeotypes, clays, polymetal oxides, sulfated zirconia, heteropolyacids, and Nafion.

It is indeed evident that the impressive advances in Friedel–Crafts acylation (as well as in other fundamental chemical reactions) observed in the past few decades are mainly due to the enormous interdisciplinary growth in catalyst design and preparation science that "… offers the possibility for selecting a catalyst with the adequate acid strength for almost any particular acid-catalyzed reaction" (A. Corma, *Chem. Rev.*, 1995, *95*, 559–614).

We want to express our hope that this book may give precious and useful information to all researchers involved in aromatic ketone preparation at both academic and industrial levels.

The authors

Giovanni Sartori was born in Casal-maggiore, Italy, in 1944. He studied chemistry at the University of Parma where he obtained his laureate degree under the guidance of Prof. Giuseppe Casnati in 1971. In 1972 he began his career at the University of Parma, working in the group of Prof. Casnati on the regio- and stereoselective functionalization of ambidental reagents. He was promoted to associate professor of organic chemistry in 1985 and full professor in 1994. He is now head of the Clean Synthetic Methodologies Groups of the University of Parma. His research interests include all aspects of heterogeneous catalysis applied to fine chemical and pharmaceutical production with particular attention to the preparation and use of supported organic (chiral) catalysts. A special area of intensive interest is also represented by the development of new ecocompatible synthetic methods mainly based on the exploitation of solvent-free and multicomponent reactions. He has published about 160 original studies and six chapters of books and has filed approximately 20 patents.

Raimondo Maggi was born in Parma, Italy, in 1963. He graduated at the University of Parma in 1989 (under the guidance of Prof. Giuseppe Casnati), and he received his PhD in organic chemistry (under the direction of Prof. Giovanni Sartori) from the same university in 1992. In 1995 he carried out his postdoctoral research with Prof. Manfred Schlosser at Lausanne University. He began his career at the University of Parma in 1997 as a researcher, and in 2002 he was promoted to associate professor of organic chemistry. His research interests include the preparation and use of heterogeneous (chiral) catalysts for the environmentally friendly synthesis of fine chemicals and pharmaceuticals. He has published about 100 original studies and six book chapters.

chapter 1

Introduction and scope

The first mention of a metal-promoted aromatic acylation appeared in 1873 in a preliminary communication by Grucarevic and Merz,[1] closely followed by a note of Zincke,[2] who described the reaction of benzoyl chloride with metals such as copper, silver, and zinc in benzene with the goal of producing benzil **1** (Scheme 1.1).

The metal was utilized to react with the liberated chlorine, giving the corresponding chloride. However, benzophenone **2** instead of benzil **1** was obtained as the main product.

The preparation of ketones from aromatic hydrocarbons and acyl chlorides in the presence of zinc metal or zinc oxide was reported by Grucareviz and Merz in the same year,[3] and by Doebner and Stackman in 1876.[4] Despite all authors having recognized the presence of zinc chloride in the final reaction mixture, no mention was made of the possible role of a metal halide as catalyst in the reaction.

It was only in 1877 that Charles Friedel (Strasbourg, France; Figure 1.1) and James Mason Crafts (Boston, Massachusetts; Figure 1.2) published together a paper entitled "Sur une nouvelle méthode générale de synthèse d'hydrocarbures, d'acétones, etc.," that gave a clear statement on the crucial role of the metal halide as catalyst in the alkylation as well as acylation of aromatic compounds.[5]

Here we report the translation made by G. A. Olah of the crucial Friedel and Crafts sentence:[6] "This last circumstance led us to discover whether the principal reaction, instead of being due to the metal as we had originally supposed, should really be attributed to the metal chloride."

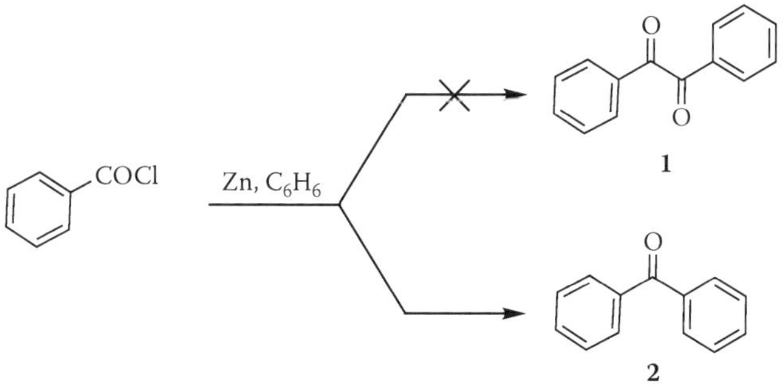

Scheme 1.1

Figure 1.2 James Mason Crafts. (From

Figure 1.1 Charles Friedel. (From http://   http://www.brynmawr.edu. With per-
www.brynmawr.edu. With permission.)   mission.)

Successively, Friedel and Crafts studied the generality and the limitations of the new synthetic method. They found that the reaction could be
successfully applied to a large number of aromatic compounds, as well as
alkyl and acyl chlorides or anhydrides in the presence of chlorides of certain
metals such as aluminum, zinc, and iron. A mechanistic hypothesis was
postulated on the basis of the possible existence of an intermediate compound **3** formed between benzene and aluminum chloride (Scheme 1.2).[7]

This intermediate would react with the electrophilic reagent, giving
the substitution product and restoring the catalyst.

A great number of theoretical and applied studies were necessary to
obtain information and proofs needed to formulate the true reaction mechanism and to establish the reactivity, and the orientation rules of that are
now generally known as the Friedel–Crafts reaction in all chemical areas.

Today, at about 130 years from its discovery, an impressive number of
scientific papers and industrial patents have been produced that deal with
Friedel–Crafts chemistry.

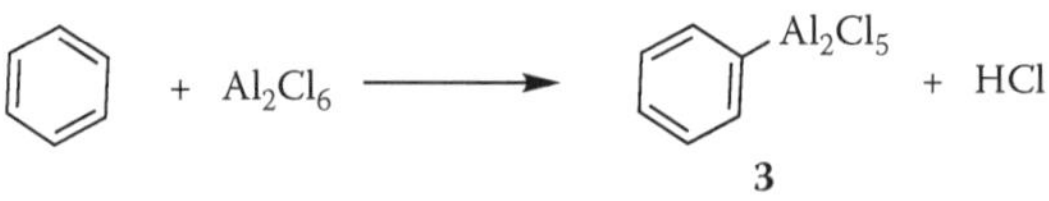

Scheme 1.2

Figure 1.3 Nobel laureate Professor George Andrew Olah. (From http://college .usc.edu/faculty/faculty1003584.html. With permission.)

From 1963 to 1965 Professor George Andrew Olah (Nobel Prize in Chemistry in 1994; Figure 1.3) edited and, in part, wrote a comprehensive treatise entitled *Friedel–Crafts and Related Reactions*. This monograph was published in four volumes covering more than 5,000 pages (with about 20,000 references). In 1972 Professor Olah published a book on the same topic entitled *Friedel–Crafts Chemistry*. The book comprises reprints of the five general review chapters written in the first version, with additions to bring them up to date until the middle of 1972. New chapters discuss general advances achieved in the field for the period 1965–1972 and the mechanistic aspects of the reactions.

Concerning, in particular, the Friedel–Crafts acylation, the reaction quickly became a fundamental pillar of synthetic organic chemistry at both the academic and industrial levels. The great interest in electrophilic acylation studies and in the optimization of preparative processes was spurred by the considerable practical value of aromatic ketone products. In fact, these compounds constitute fundamental intermediates in the pharmaceutical, fragrance, flavor, dye, and agrochemical industry.[8–12]

To mention some examples, benzoylation of xylenes was particularly studied because of the use of dimethylbenzophenones as UV-light stabilizers in plastics.[13] Moreover, 2-acetyl-6-methoxynaphthalene represents the precursor to the antirheumatic Naproxen.[14] *ortho*- and *para*-Hydroxyacetophenones are widely used for the synthesis of aspirin and paracetamol (4-acetamidophenol), respectively.[15] *ortho*-Hydroxyacetophenone is also a key intermediate in the production of 4-hydroxycoumarin and warfarin, both used as anticoagulant drugs in the therapy of thrombotic diseases,[16] and it is also employed in the synthesis of flavonones.[17,18]

Conventionally, electrophilic acylation is catalyzed by Lewis acids (such as zinc chloride, aluminum trichloride, iron trichloride, tin tetrachloride, and titanium tetrachloride) or strong protic acids (such as hydrofluoric acid and sulfuric acid). In particular, the use of metal halides causes problems associated with the strong complex formed between the ketone product and the metal halide itself, which provokes the use of more than stoichiometric amounts of catalyst. The workup commonly requires hydrolysis of the complex, leading to the loss of the catalyst and giving large amounts of corrosive waste streams. For these reasons, during the past decades, the setting up of more efficient and ecocompatible Friedel–Crafts acylation processes has become a fundamental goal of the general "green revolution" that has spread in all fields of synthetic chemistry[19] through a revision of the preparation methods mainly based on the exploitation of new and increasingly efficient catalysts.[20]

Great efforts have been made to achieve the goal of making the Lewis- or protic-acid role in Friedel–Crafts acylation catalytic. It is indeed well recognized that catalysis plays a crucial role toward the development of clean technologies. Efficient catalysis means minimizing energy consumption, achieving high yield and selectivity, using green solvents, or, better, avoiding use of any solvent.

Among the new synthetic methodologies recently studied and exploited, the use of ionic liquids as solvent-catalysts to perform more environmentally friendly Friedel–Crafts acylations has been developed.[21] In contrast to volatile organic solvents, ionic liquids have no measurable vapor pressure, and, therefore, there is no loss of solvent through evaporation. Moreover, ionic liquids can be easily recovered, cleaned, and reused for different runs.

Moreover, Friedel–Crafts reactions have been applied as a crucial step in multicomponent[22] as well as multistep cascade reactions.[23] These approaches have considerable importance from the economic and environmental point of view since they allow the building of target compounds with great structural complexity through a one-pot process.

A review has been published by D. E. Pearson and C. A. Buehler showing numerous examples of electrophilic aromatic acylations that proceed satisfactorily with a small amount of catalyst.[24]

A large variety of catalysts are effective in reactions between activated aromatic or heteroaromatic substrates and acyl chlorides or anhydrides. They include metals such as iron and copper, nonmetals such as iodine, metal halides such as ferric chloride and zinc chloride, and acids such as hydroiodic acid.

More recently, interesting results have been achieved, such as the use of catalytic amounts of lanthanide trifluoromethane sulfonates (triflates) alone[25] or microencapsulated on polyacrylonitrile[26] as reusable catalysts; moreover ecocompatible methods have been set up based on the use of graphite as a solid catalyst[27] or even exploiting the possibility to perform the reaction without any catalyst.[28]

Maximum effort has been directed toward the use of solid acid catalysts. In fact, heterogeneous catalysts can be easily separated from the reaction mixture and reused; they are generally not corrosive and do not produce problematic side products. Different classes of materials have been studied and utilized as heterogeneous catalysts for Friedel–Crafts acylations: these include zeolites (acid treated), metal oxides, and heteropoly acids already utilized in hydrocarbon reactions.[29] Moreover, the application of clays, perfluorinated resinsulfonic acids, and supported (fluoro) sulfonic acids, mainly exploited in the production of fine chemicals, are the subject of intensive studies in this area.

Owing to the great interest in the argument, minireviews have been published on the use of solid catalysts in Friedel–Crafts acylation. Kouwenhoven and van Bekkum, in a chapter of the *Handbook of Heterogeneous Catalysis*, faced the basic problem of the use of zeolites in the reaction.[30] A further essential overview of the same argument was reported by Métivier in *Fine Chemicals through Heterogeneous Catalysis*.[31] Furthermore, Bezouhanova described the synthetic aspects of the zeolite-catalyzed preparation of aromatic ketones.[32]

The present monograph reviews the most important studies in the Friedel–Crafts acylation reaction published during the past three decades, with particular emphasis on the improvement achieved by the synthetic point of view. According to the modern concept of process efficiency, not only product yield and selectivity but also the use of solvents and their green nature, energetic economy, and catalyst activity are considered as positive parameters in comments on the mentioned studies.

The whole argument is organized in four sections: stoichiometric acylations, catalytic homogeneous acylations, catalytic heterogeneous acylations, and phenol acylations. It is structured according to the role played by the catalyst in the activation of reagents as well as in the different modes of regioselectivity encountered in the acylation of arenes, aromatic ethers, and phenols.

Even if the great majority of the papers reviewed concern reactions carried out in batch conditions, we have particularly stressed the results obtained in gas-phase or under continuous fixed-bed reaction conditions, where the continuous removal of products and poisons from the catalyst surface enhances performance.

References

1. Grucarevic, S. and Merz, V. 1873. Zur synthese der ketone. *Chem. Ber.* 6: 60.
2. Zincke, T. 1873. Uber die einwirkung von zink auf gemische von aromatischen haloidverbindungen mit kohlenwasserstoffen. *Chem. Ber.* 6: 137–139.
3. Grucarevic, S. and Merz, V. 1873. Ketone aus aromatischen kohlenwasserstoffen und säurechloriden. *Chem. Ber.* 6: 1238–1246.

4. Doebner, D. and Stackman, W. 1876. Ueber die einwirkung des benzo-trichlorids auf phenol. *Chem. Ber.* 9: 1918–1920.

5. Friedel, C. and Crafts, J. M. 1877. Sur une nouvelle méthode générale de syn-thèse d'hydrocarbures, d'acétones, etc. *C. R.* 84: 1450.

6. Olah, G. A. 1964. *Friedel–Crafts and related reactions*. London: Interscience, 1: 12.

7. Friedel, C. and Crafts, J. M. 1878. Synthése du cresol et de l'acide benzoique. *Bull. Chem. Soc. Fr.* 29: 242.

8. Franck, H. G. 1988. *Industrial aromatic chemistry*. Berlin: Springer.

9. Horsley, J. A. 1997. Producing bulk and fine chemicals using solid acids. *Chemtech* October: 45–49.

10. Bauer, K., Garbe, D., and Surberg, H. 1990. *Common fragrance and flavor materials*. Weinheim: WHC Vertagsgesellschaft.

11. Buu-Hoï, N. P., Xuong, N. D., and Suu, V. T. 1961. Spasmolytic ketones derived from benzofuran. *C. R. Acad. Sci.* 253: 1075–1076.

12. Murti, V. A., Bhandari, K., Jain, P. C., and Anand, N. 1989. Agents acting on the central nervous system. Part XXXV. 1-(Benzofuran-2/3-yl)-2/3-amino-alkan-1-ones/-1-ols. *Ind. J. Chem.* 28B: 385–390.

13. Yadav, G. D., Asthana, N. S., and Kamble, V. S. 2003. Friedel–Crafts benzoyla-tion of *p*-xylene over clay-supported catalysts: novelty of cesium-substituted dodecatungstophosphoric acid on K-10 clay. *Appl. Catal. A: General* 240: 53–69.

14. Villa, C. G. M. and Panossian, S. P. 1992. In *Chirality in industry*, eds. A. N. Collins, G. N. Sheldrake, and J. Crosby, 303. New York: John Wiley & Sons.

15. Fritch, J. R., Fruchey, O. S., and Horlenko, T. 1990. Production of acetamino-phen. U.S. Patent 4,954,652.

16. Uwaydah, I. M., Aslam, M., Brown, C. H., Fitzhenry, S. R., and Mcdonough, J. A. 1997. Syntheses based on 2-hydroxyacetophenone. U.S. Patent 5,696,274.

17. Climent, M. J., Corma, A., Iborra, S., and Primo, J. 1995. Base catalysis for fine chemicals production: Claisen–Schmidt condensation on zeolites and hydro-talcites for the production of chalcones and flavanones of pharmaceutical interest. *J. Catal.* 151: 60–66.

18. Drexler, M. T. and Amiridis, M. D. 2003. The effect of solvents on the hetero-geneous synthesis of flavanone over MgO. *J. Catal.* 214: 136–145.

19. Sheldon, R. A. and Downing, R. S. 1999. Heterogeneous catalytic transfor-mations for environmentally friendly production. *Appl. Catal. A: Gen.* 189: 163–183.

20. Anastas, P. T., Bartlett, L. B., Kirchhoff, M. M., and Williamson, T. C. 2000. The role of catalysis in the design, development, and implementation of green chemistry. *Catal. Today* 55: 11–22.

21. Dupont, J., de Souza, R. F. and Suarez, P. A. Z. 2002. Ionic liquid (molten salt) phase organometallic catalysis. *Chem. Rev.* 102: 3667–3692.

22. Dömling, A. and Ugi I. 2000. Multicomponent reactions with isocyanides. *Angew. Chem. Int. Ed.* 39: 3168–3210.

23. Tietze, L. F. and Beifuss, U. 1993. Sequential transformations in organic chemistry: a synthetic strategy with a future. *Angew. Chem., Int. Ed. Engl.* 32: 131–163.

24. Pearson, D. E. and Buehler, C. A. 1972. Friedel–Crafts acylations with little or no catalyst. *Synthesis* 533–542.

25. Kawada, A., Mitamura, S., and Kobayashi, S. 1993. Lanthanide trifluoromethanesulfonates as reusable catalysts: catalytic Friedel–Crafts acylation. *J. Chem. Soc., Chem. Commun.* 1157–1158.
26. Kobayashi, S. and Nagayama, S. 1998. A microencapsulated Lewis acid. A new type of polymer-supported Lewis acid catalyst of wide utility in organic synthesis. *J. Am. Chem. Soc.* 120: 2985–2986.
27. Kodomari, M., Suzuki, Y., and Yoshida, K. 1997. Graphite as an effective catalyst for Friedel–Crafts acylation. *J. Chem. Soc., Chem. Commun.* 1567–1568.
28. Mil'to, V. I., Mironov, G. S., and Kopeikin, V. V. 1989. High-temperature benzoylation. *Zh. Org. Khim.* 25: 2372–2374.
29. Corma, A. 1995. Inorganic solid acids and their use in acid-catalyzed hydrocarbon reactions. *Chem. Rev.* 95: 559–614.
30. Kouwenhoven, H. W. and van Bekkum, H. 1997. Acylation of aromatics. In *Handbook of heterogeneous catalysis*, eds. G. Ertl, H. Knözinger, and J. Weitkamp, Vol. 5, 2358–2364. Weinheim: VCH.
31. Métivier, P. 2001. Friedel–Crafts acylation. In *Fine chemicals through heterogeneous catalysis*, eds. R. A. Sheldon and H. van Bekkum, 161–172. Weinheim: VCH.
32. Bezouhanova, C. P. 2002. Synthesis of aromatic ketones in the presence of zeolite catalysts. *Appl. Catal. A: Gen.* 229: 127–133.

chapter 2

Stoichiometric acylations

The classical Friedel–Crafts acylation reactions are usually carried out in the presence of a stoichiometric amount of Lewis acid catalyst and are very familiar today to all chemical research workers. In this chapter, only special examples of stoichiometric acylation will be commented. For example, reactions showing extraordinary level of regioselectivity promoted by proximity or metal template effects are described. Moreover, examples of efficient use of carboxylic acids and esters as acylating agents under soft experimental conditions in combination with ecocompatible solvents are stressed as new and practicable synthetic methods. Studies on the highly efficient multistep synthesis of polyfunctional compounds via bis-acylation and alkylation–acylation processes are commented upon, and some mechanistic details are also shown.

2.1 Acylation

The proximity effect of acyl groups covalently bounded to heteroaromatic compounds is advantageously exploited in the intramolecular regioselective electrophilic acylation of N-substituted pyrroles (Scheme 2.1).[1]

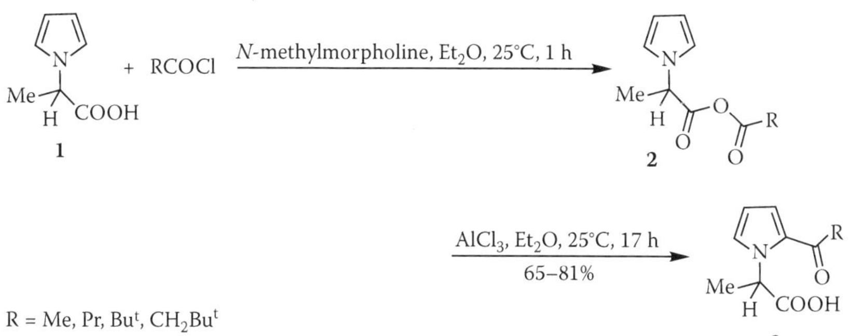

R = Me, Pr, Bu^t, CH$_2$Bu^t

Scheme 2.1

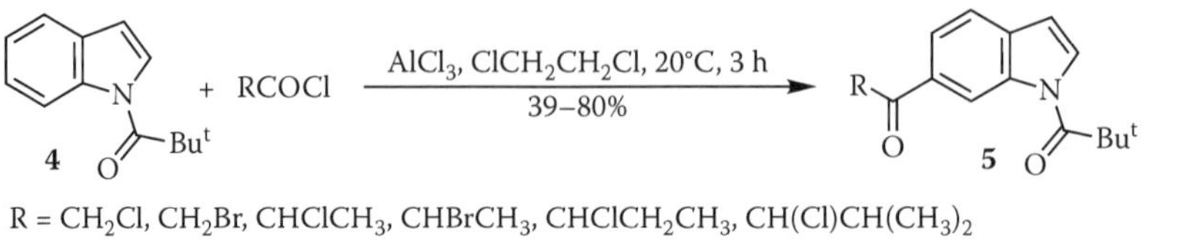

R = CH$_2$Cl, CH$_2$Br, CHClCH$_3$, CHBrCH$_3$, CHClCH$_2$CH$_3$, CH(Cl)CH(CH$_3$)$_2$

Scheme 2.2

Thus, compound **1**, easily synthesized from proline and 2,6-dimethoxytetrahydrofuran by the modified Clauson–Kaas procedure,[2] is converted into the mixed anhydrides **2** by reaction with different acyl chlorides and N-methylmorpholine. Treatment of compounds **2** with a stoichiometric amount of aluminum chloride in dry diethyl ether gives the C-2 derivatives **3** in 65%–81% yield via intramolecular acyl transfer controlled by the proximity effect.

Trace amounts of 3-acyl derivatives are also produced by conventional intermolecular acylation from another molecule of mixed anhydride **2**.

Regioselective substitution on indole is one of the most important goals in heterocyclic chemistry owing to the great importance of indoles in the preparation of biologically active products. The 2- and 3-positions of the ring are the normal sites of attack in the electrophilic substitution, including acylation because of their much more nucleophilic activity.[3]

The reaction of 1-pivaloylindole **4** with excess of α-halogenoacyl chlorides in 1,2-dichloroethane in the presence of an excess of aluminum chloride gives the unusual 6-acylation, affording 6-acylindoles **5** in 39%–80% yield as the sole reaction products (Scheme 2.2).

The use of indole salts is frequently required to achieve selective attack at the 3-position in the acylation reaction. However, this approach is not applicable to indoles bearing functional groups labile under basic conditions, and consequently, Grignard reagents or alkylzinc compounds cannot be used for the preparation of the indole salts as shown in Scheme 2.2. A second approach involves the use of N-protected indoles and requires protection–deprotection steps.

The direct and selective 3-acylation of both acid- and base-sensitive indoles **6** can be efficiently achieved using acyl chlorides and alkyl aluminum chlorides without NH protection (Scheme 2.3).[4] When acyl chlorides are used for the acylation of indoles, liberation of hydrogen chloride is unavoidable, and the reaction results in the production of tar materials mainly due to polymerization of indole. Alkylaluminum compounds are utilized to scavenge the acid in the acylation with acyl chlorides. The reaction proceeds cleanly to give 3-acylindoles **7** without any by-product formation. The use of triethylaluminum leads to lower yields, probably because a nucleophilic attack of the alkyl group to acyl chloride occurs competitively. As expected, the use of aluminum

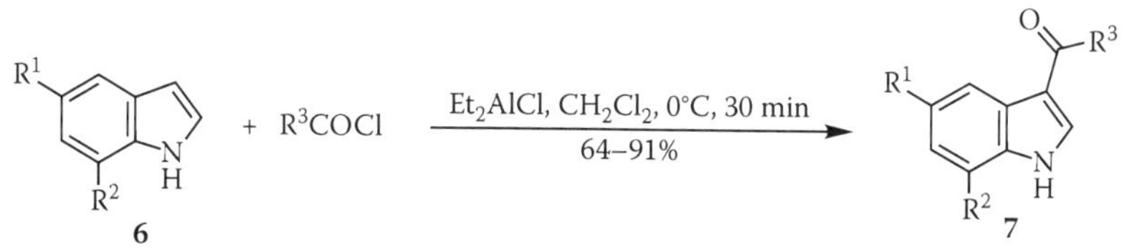

R^1 = H, OMe, CN, CO$_2$Et
R^2 = H, Me, NO$_2$
R^3 = Me, Et, Pr^i, Bu^t, CH=CHCH$_3$, CH=C(CH$_3$)$_2$, Ph, 2-furyl, CO$_2$Et

Scheme 2.3

chloride results in decomposition and oligomerization of indole due to its strong Lewis acidity.

The aluminum chloride-promoted Friedel–Crafts acylation can also be successfully performed in ionic media. An easy-to-handle ionic liquid is that prepared from a mixture of one to two equivalents of aluminum chloride and 1-ethyl-3-methylimidazolium chloride or iodide (EtMeimCl-AlCl$_3$ and EtMeimI-AlCl$_3$, respectively). These solvent-catalysts can be utilized in the easy and efficient acylation of unprotected indoles with a variety of acyl chlorides to give C-selective acylation at the 3-position in 38%–87% yield.[5]

The acylation of arenes with succinic and phthalic anhydrides in a chloroaluminate ionic liquid gives the corresponding ketoacids in good yields (50%–95%). The system follows a typical electrophilic behavior, and lower yields are achieved with electron-poor aromatics.[6]

Good synthetic results have been achieved in the acylation of ferrocene with acetic anhydride (AAN) in EtMeimCl-AlCl$_3$/toluene mixture. The monoacetylated product can be obtained in 80%–96% yield. It is significant that no acetylation of toluene is observed under these conditions.

Unfortunately, the workups employed in all the aforementioned synthetic methods involve quenching with acid water, and consequently the solvent-catalysts are destroyed.[7]

A quite interesting direct alkylacylation of arenes with acyl halides and alkanes in the presence of aluminum halides can be performed. The reaction involves the highly electrophilic derivatives **8**, prepared by mixing acid halides and aluminum halides (Scheme 2.4). Compounds **8** can easily generate carbenium ions from saturated hydrocarbons and can be consequently utilized in the effective alkylacylation of benzene and bromobenzene, yielding para-alkylated aromatic ketones **9** through a cascade reaction.[8] The process suffers from some synthetic limitations being efficiently applied only to benzene or bromobenzene and acetyl or propanoyl chlorides. The proposed mechanism involves the generation of a carbenium ion from the alkane, which undergoes alkylation of the arene with subsequent acylation of the alkylated aromatic compound

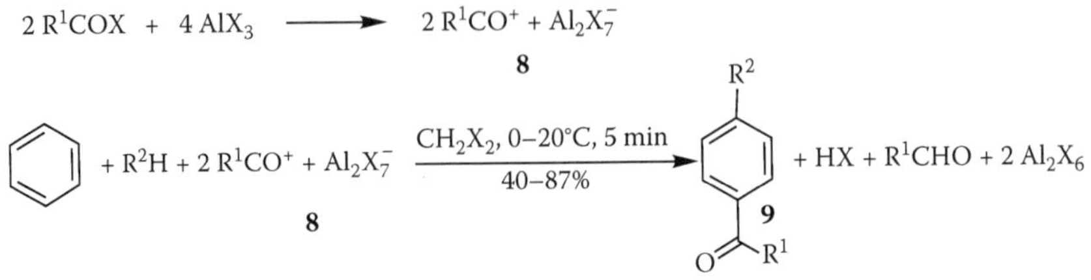

$$2\,R^1COX + 4\,AlX_3 \longrightarrow \underset{\mathbf{8}}{2\,R^1CO^+ + Al_2X_7^-}$$

$$\underset{\mathbf{8}}{\bigcirc + R^2H + 2\,R^1CO^+ + Al_2X_7^-} \xrightarrow[40\text{–}87\%]{CH_2X_2,\ 0\text{–}20°C,\ 5\ min} \underset{\mathbf{9}}{\bigcirc} + HX + R^1CHO + 2\,Al_2X_6$$

R^1 = Me, Pr, Ph
R^2 = Bu, Bu^i, C_5H_{11}, Me$_2$CH(CH$_2$)$_2$, Me$_2$CH CH(Me)CH$_2$
X = Cl, Br

Scheme 2.4

$$R^1COX + Al_2X_6 \rightleftharpoons \underset{\mathbf{8}}{R^1CO^+ + Al_2X_7^-}$$

$$R^2H + \underset{\mathbf{8}}{R^1CO^+ + Al_2X_7^-} \rightleftharpoons R^{2+} + Al_2X_7^- + R^1CHO$$

$$R^3C_6H_5 + R^{2+} + Al_2X_7^- \rightleftharpoons R^3C_6H_4R^2 + Al_2X_6 + HX$$

$$R^3C_6H_4R^2 + \underset{\mathbf{8}}{R^1CO^+ + Al_2X_7^-} \rightleftharpoons R^3C_6H_3(R^2)COR^1 + Al_2X_6 + HX$$

R^1 = Me, Pr, Ph
R^2 = Bu^i, C_5H_{11}, Me$_2$CH(CH$_2$)$_2$, Me$_2$CH(CH$_2$)$_3$
R^3 = H, Br
X = Cl, Br

Scheme 2.5

(Scheme 2.5). The metal-free acylation of aromatics and heteroaromatics with carboxylic acids can be performed by using excess trifluoroacetic anhydride (TFAA) and 85% phosphoric acid as catalyst (Table 2.1).[9] The reaction affords alkyl aryl ketones and heteroaryl ketones in generally good yields (23%–82%). Furane, thiophene, anisole, pentamethylbenzene, and ferrocene react with a large variety of carboxylic acids, including long-chain ones (i.e., dodecanoic acid). The reaction initially involves the mixed alkanoic–TFAA (acyl trifluoroacetates)[10] that can be isolated and characterized by IR spectroscopy. The acylation can be performed also in the absence of 85% phosphoric acid; however, in these cases it requires significantly longer reaction times and gives lower yields.

An approach to the more atom-efficient and less wasteful Friedel–Crafts acylation process is based on the in situ formation of acyl bis(trifluoroacetyl) phosphates showing a much greater degree of activation at the covalent

Table 2.1 Acylation of (hetero)aromatics with carboxylic acids promoted by TFAA–phosphoric acid mixture

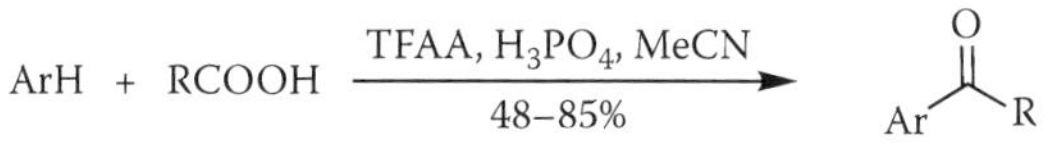

$$\text{ArH} + \text{RCOOH} \xrightarrow[\text{48–85\%}]{\text{TFAA, H}_3\text{PO}_4\text{, MeCN}} \underset{\text{Ar}}{\overset{\text{O}}{\|}}\text{R}$$

ArH	R	T (°C)	t (min)	ArCOR	Yield (%)
	Pr	20	30	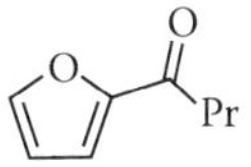	75
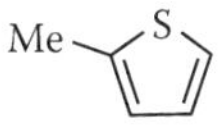	Bu	20	4	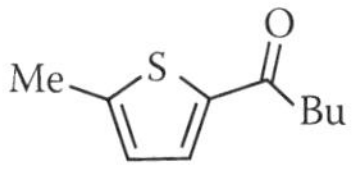	78
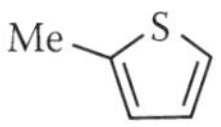	$C_{11}H_{23}$	50	8	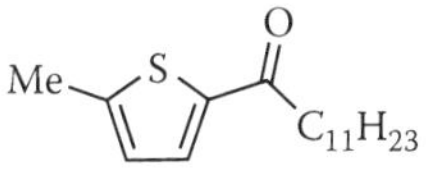	81
	Me	Reflux	20	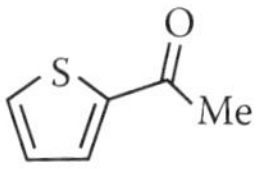	77
	$C_{11}H_{23}$	50	120	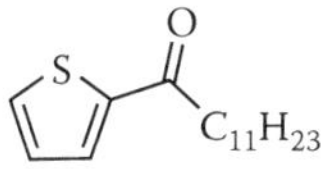	82
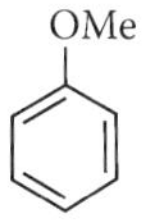	Me	Reflux	80	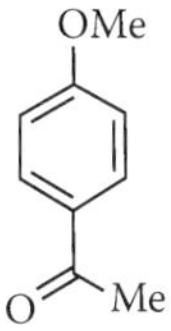	66
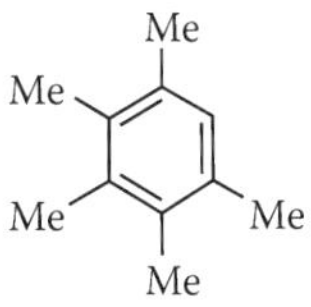	Me	Reflux	900	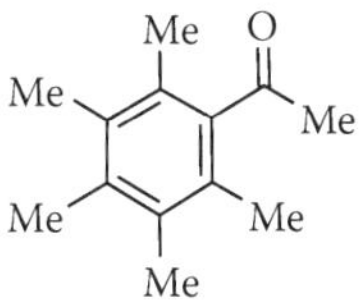	23
	Me	20	30	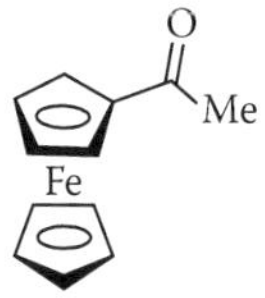	70

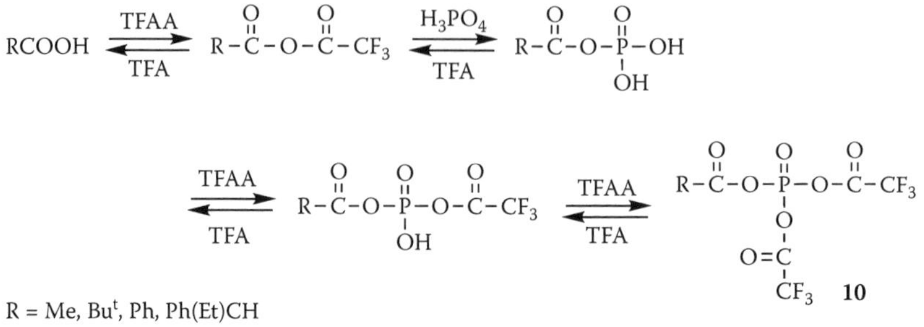

$$RCOOH \underset{TFA}{\overset{TFAA}{\rightleftharpoons}} R-\overset{O}{\overset{||}{C}}-O-\overset{O}{\overset{||}{C}}-CF_3 \underset{TFA}{\overset{H_3PO_4}{\rightleftharpoons}} R-\overset{O}{\overset{||}{C}}-O-\overset{O}{\underset{OH}{\overset{||}{P}}}-OH$$

$$\underset{TFA}{\overset{TFAA}{\rightleftharpoons}} R-\overset{O}{\overset{||}{C}}-O-\overset{O}{\underset{OH}{\overset{||}{P}}}-O-\overset{O}{\overset{||}{C}}-CF_3 \underset{TFA}{\overset{TFAA}{\rightleftharpoons}} R-\overset{O}{\overset{||}{C}}-O-\overset{O}{\underset{\underset{CF_3}{\overset{|}{O=C}}}{\overset{||}{P}}}-O-\overset{O}{\overset{||}{C}}-CF_3 \quad \mathbf{10}$$

R = Me, Bu^t, Ph, Ph(Et)CH

Scheme 2.6

bond formation state and consequently requiring milder acids.[11] Detailed ^{19}F and ^{31}P NMR spectroscopic studies allow the observation and characterization of the real active catalytic agent. The in situ reaction of TFAA with the appropriate carboxylic acid and phosphoric acid at 10°C is shown to afford the acyl bis(trifluoroacetyl)phosphate **10** as the most abundant actual acylating agent (Scheme 2.6). The reaction occurs in high or quantitative yield with electron-rich aromatic substrates carrying methoxy and alkyl groups, with exclusive formation of the para isomer.

A further application of this approach is the synthesis of 1,2-diaryl-1-ethanones **12** in good to excellent yields by reaction of aryl acetic acids **11** with arenes in the presence of 85% phosphoric acid without solvent at 50°C for 30–180 min (Scheme 2.7).[12] The reaction tolerates a variety of substituents, and groups such as methoxy and thiomethoxy are found to be highly effective.

The method based on the in situ production of acyltrifluoroacetates represents a useful alternative to the classic Friedel–Crafts acylation and, in some instances, it can be applied for large-scale preparation.[13] It must be underlined that the atom efficiency of the acyltrifluoroacetate-based method is augmented by recovery of the spent TFAA as trifluoroacetic acid TFA and conversion of this back to TFAA.

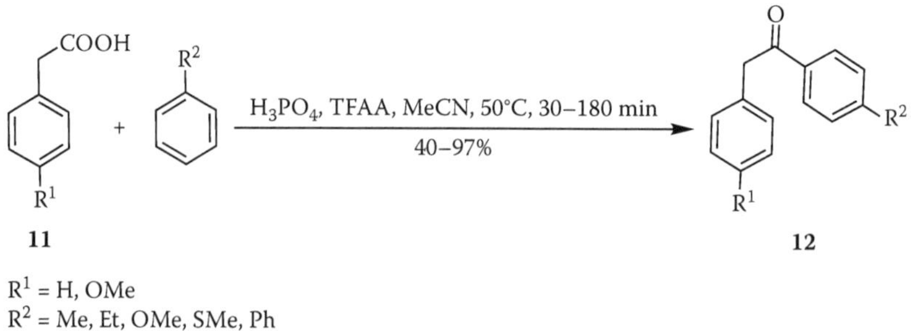

R^1 = H, OMe
R^2 = Me, Et, OMe, SMe, Ph

Scheme 2.7

Table 2.2 Benzoylation of aromatics with methyl benzoate in the presence of triflic acid

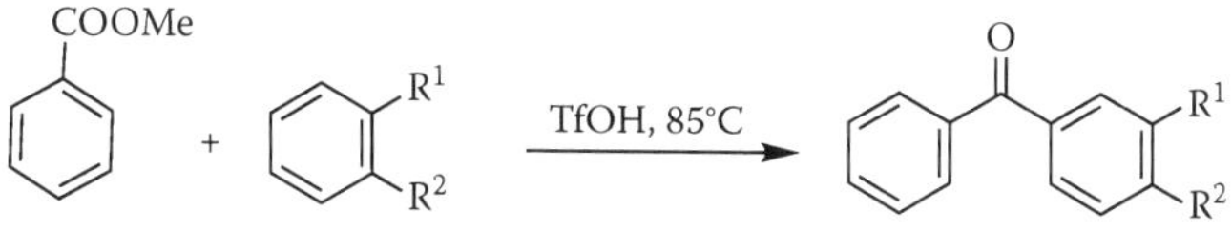

R^1	R^2	t (h)	Yield (%)
H	H	8	75
H	Cl	1	78
H	F	1	85
H	Me	0.5	70
H	OMe	2	93
Me	Me	2	76
NO_2	H	8	82
CF_3	H	2	84

The direct acylation of aromatics with methyl benzoate can be achieved by using triflic acid as Brönsted acid catalyst.[14] The yield of benzophenones obtained ranges from 70% to 93%. Even highly deactivated nitrobenzene and benzotrifluoride show good reactivity and give the corresponding meta-substituted benzophenones in high yield (Table 2.2). Benzoic acid itself does not react with benzene under similar reaction conditions. A possible mechanism of acylation involves ester protonation, giving activated mono- and dicationic species **13, 14,** and **15** (Scheme 2.8).

Friedel–Crafts acylation utilizing β-lactams **16** as acylating agents in the presence of triflic acid is reported as an efficient method to produce β-aminoaromatic ketone derivatives **17** (Scheme 2.9).[15] The reaction occurs under very mild conditions such as 0°C room temperature for 15 min for both activated and deactivated aromatic compounds. As an example, chloro- and bromobenzene give the corresponding para-acylated products in 92% yield. Naphthalene reacts with N-(2,2,2-trichloroethoxycarbonyl)-protected β-lactam **16** ($R^1 = CO_2CH_2CCl_3$), giving a 1/0.78 mixture of 1- and 2-substituted isomers.

Different patents have been registered concerning the preparation of specific aryl ketones by Friedel–Crafts acylation of convenient aromatic substrates in the presence of variable amounts of both Brönsted and Lewis acids. As an example, a large number of aryl ketones was efficiently synthesized through acylation of arenes with acyl chlorides in methylene chloride and in the presence of a mixture of aluminum chloride and lithium chloride as the catalyst.[16] Some synthetic results are reported in Table 2.3.

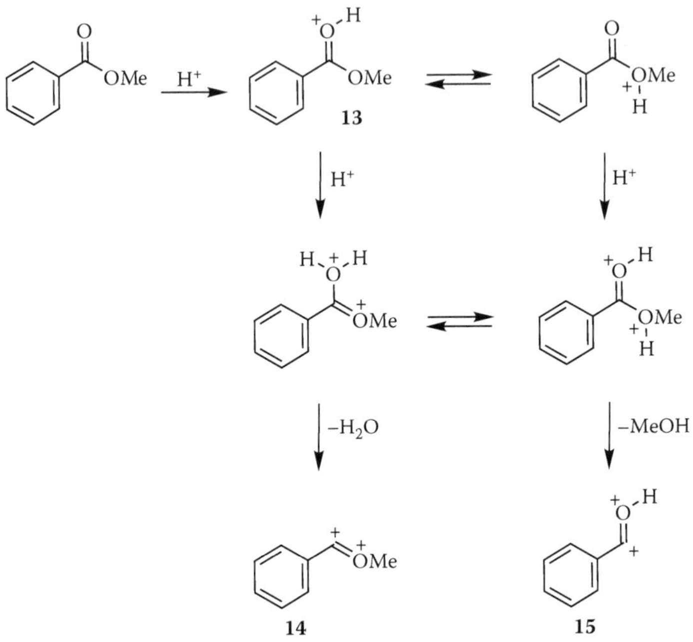

Scheme 2.8

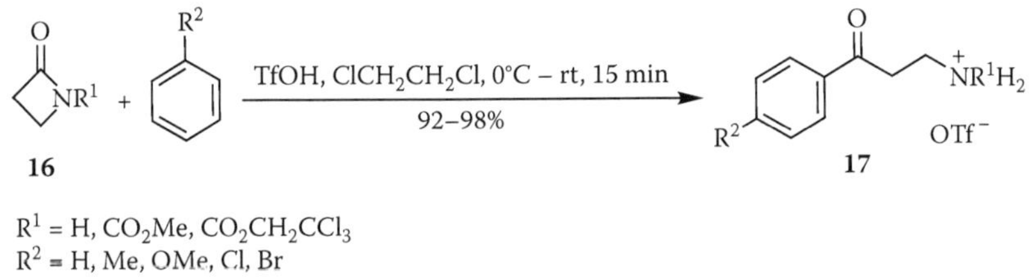

R^1 = H, CO$_2$Me, CO$_2$CH$_2$CCl$_3$
R^2 = H, Me, OMe, Cl, Br

Scheme 2.9

The acetylation of isobutylbenzene with AAN catalyzed by boron trifluoride can be performed in anhydrous sulfur dioxide at −40°C and heating the autoclave container to 22°C for about 2 h.[17] The process can be applied to different aromatic compounds, including alkylarenes, naphthalenes, diphenyl ethers, and thioethers.

2,4-Dihydroxyacetophenones **20** are synthesized with satisfactory yield through a multistep process (Scheme 2.10).[18,19] Compounds **19** carrying an alkyl chain between the two methoxy groups are synthesized by the well-known method based on the selective deprotonation of **18** at the C-2 position.[20] The electrophilic acetylation of compounds **19** can be performed under mild conditions with acetic acid in the presence of hydrobromic acid

Table 2.3 Aluminum chloride–lithium chloride catalyzed acylation of aromatics with acyl chlorides in 1,2-dichloroethane

Substrate	Acylating agent	Conditions	Product	Yield (%)
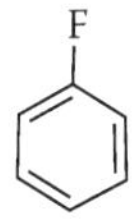	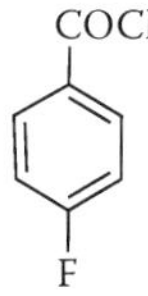 COCl	−15°C, 1 h; 0°C, 3 h; rt, overnight	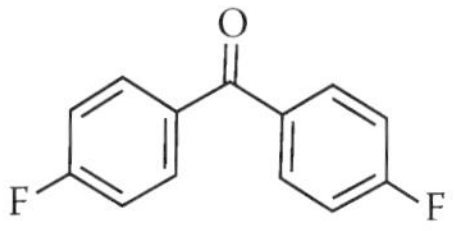	95
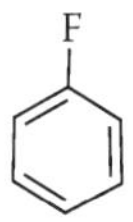	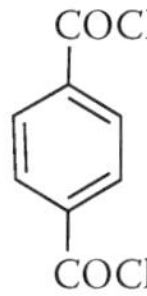 COCl COCl	0°C, overnight	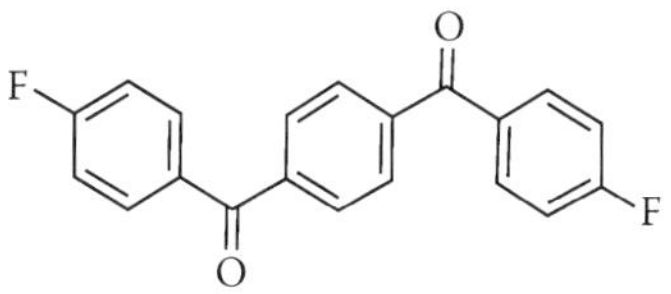	87
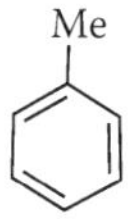	MeCOCl	−15°C, 1 h; rt, overnight	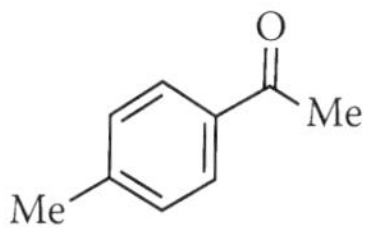	95
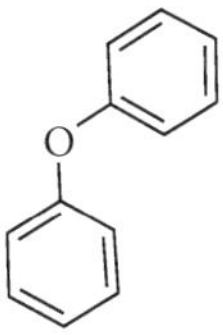	MeCOCl	−15°C, 1 h; rt, overnight	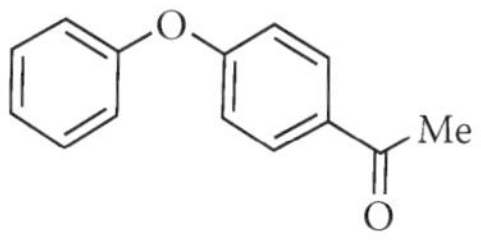	90
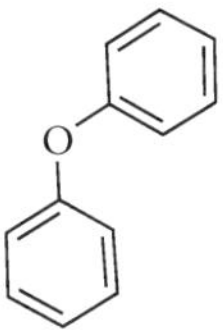	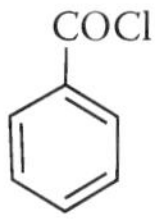 COCl	−15°C, 1 h; rt, overnight	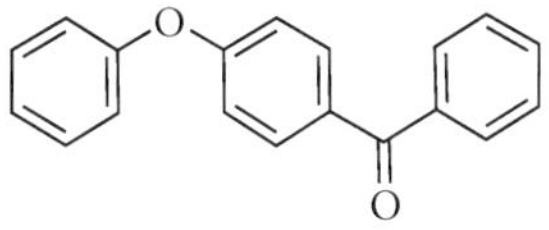	94
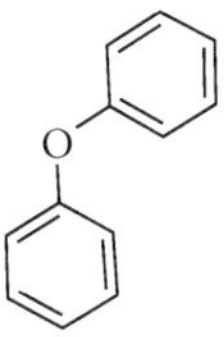	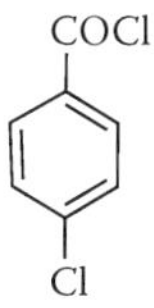 COCl Cl	−15°C, 1 h; rt, overnight	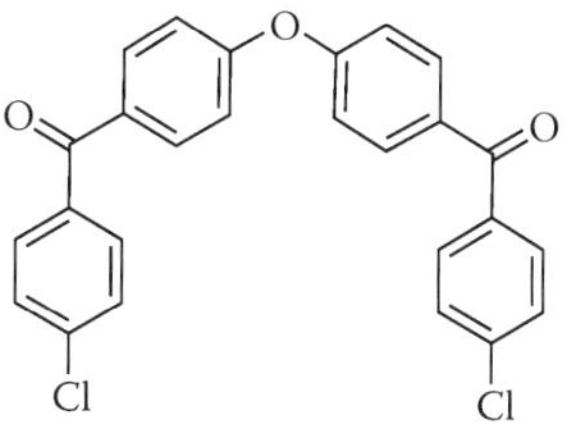	97
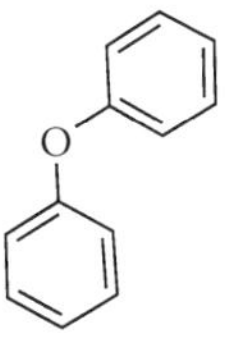	COCl$_2$	−15°C, 1 h; rt, 68 h	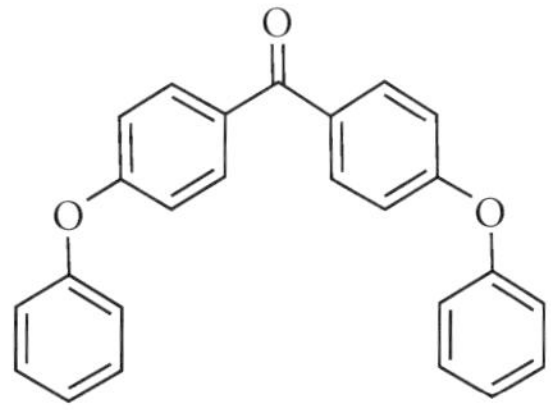	60

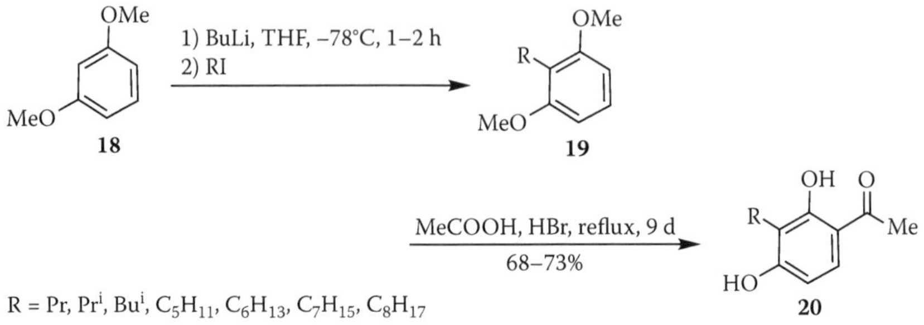

R = Pr, Pr^i, Bu^i, C$_5$H$_{11}$, C$_6$H$_{13}$, C$_7$H$_{15}$, C$_8$H$_{17}$

Scheme 2.10

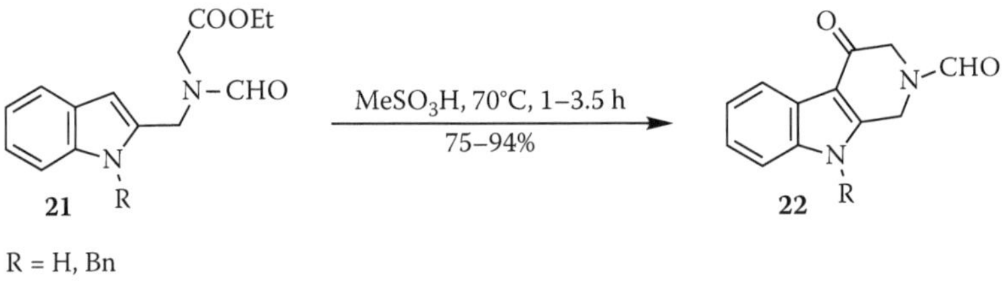

R = H, Bn

Scheme 2.11

as protic acid catalyst. Moreover, hydrobromic acid plays a crucial role well in the demethylation of compounds **19**. The final 2,4-dihydroxyacetophe-nones **20** can be obtained in 68%–73% yield.

Indole derivative **21** (Scheme 2.11, R = H) undergoes highly selective cycloacylation, affording compound **22** (R = H) in 75% yield, which is a synthon for carboline preparation.[21] Moreover, better results (94% yield) are achieved by using synthon **21** (R = Bn) with both nitrogen atoms pro-tected (benzyl protective group for the heterocyclic nitrogen and formyl group for the side chain nitrogen).

2.2   Cycloacylation

Friedel–Crafts cycloacylation represents the most typical route to poly-cyclic ketones (Scheme 2.12). The topic has been extensively examined in the basic survey by G. A. Olah[22] in which are reported the results of 461 studies showing that the variously substituted arylaliphatic carboxylic

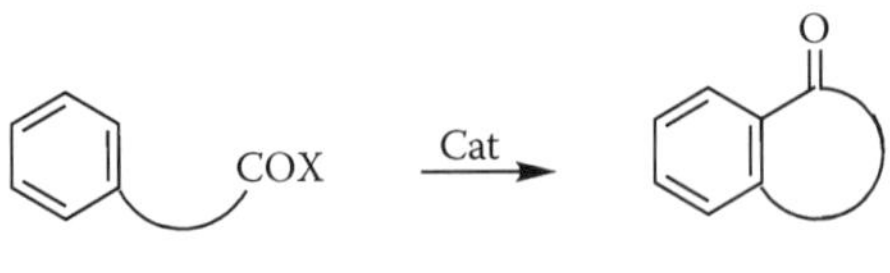

Scheme 2.12

acids or derivatives such as acyl chlorides and esters undergo ring closure in the presence of a suitable condensing agent such as sulfuric acid, hydrogen fluoride, polyphosphoric acid, or typical Lewis acids such as aluminum chloride or tin tetrachloride.

The reaction is affected by several factors, including the nature of the cycloacylating functional group, the Lewis or protic acid, the nature and position of substituents on the aromatic ring, and the size of the ring to be formed. In general, experimental conditions can be set up in order to efficiently synthesize cyclic aromatic and heteroaromatic ketones.

Good yields of cyclic aryl ketones are achieved with arenes, provided they are not very electron deficient. For example, the cycloacylation of arylpropanoic acids can be performed in the presence of scandium triflate as Lewis acid to give indanones in good yields with electron-rich aromatic substrates, whereas lower yields are obtained with very electron-deficient ones.[23] The reaction can also be performed in the presence of protic acid; thus, 3-(3,5-dibromophenyl)propanoic acid is directly converted into 2,4-dibromoindanone by treatment with polyphosphoric acid in 53% yield. The corresponding acyl chloride can be converted into the same product upon treatment with iron trichloride only in 45% yield. The efficiency of this process can be improved by using more acidic promoters such as chlorosulfonic acid.[24] The reaction occurs at 0°C for short times, and a series of electron-poor indanones **23** can be synthesized with satisfactory yields (Scheme 2.13). 6-Trifluoromethyl indanone is obtained in only 23% yield; however, the only published synthesis of this compound requires a multistep complex process.[25] It must be remarked that, in the case of highly electron-rich arenes, sulfonylation and chlorosulfonylation of the aromatic ring competes with cyclization.

In situ production of triphenyl phosphonium anhydrides as triflate salts **25** (Scheme 2.14) is claimed to be the crucial step for the cycloacylation of arylalkanoic acids in the presence of triphenyl phosphine oxide and triflic anhydride. The reaction proceeds in satisfactory to high yields under mild conditions.[26] Compounds **25** are commonly used directly in situ owing to their very hygroscopic nature and are produced by reaction

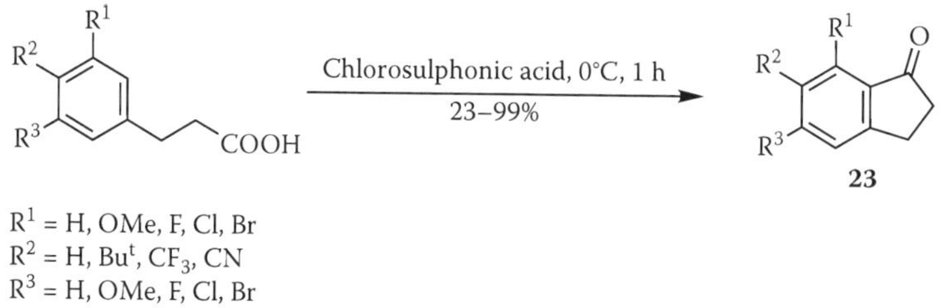

R^1 = H, OMe, F, Cl, Br
R^2 = H, But, CF_3, CN
R^3 = H, OMe, F, Cl, Br

Scheme 2.13

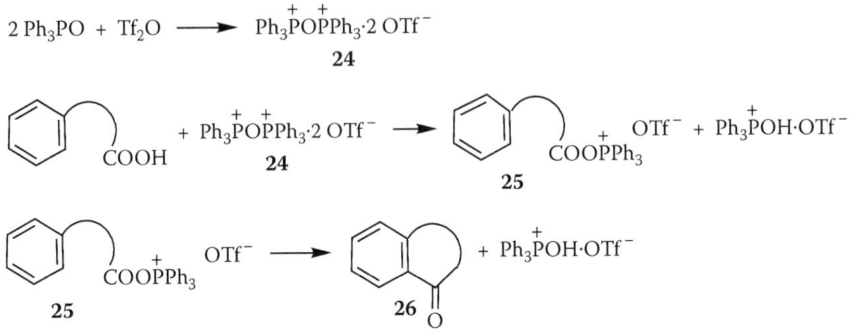

$$2\ Ph_3PO\ +\ Tf_2O\ \longrightarrow\ Ph_3\overset{+}{P}O\overset{+}{P}Ph_3\cdot2\ OTf^-$$
$$\mathbf{24}$$

Scheme 2.14

of arylalkanoic acids with the phosphonium anhydride **24**; intermediates **25**, carrying an activated carboxylic group, undergo easy cycloacylation to cycloalkanones **26** in good yields. This C-acylation is generally milder and cleaner than the traditional polyphosphoric-acid-catalyzed procedure.[27]

The synthetic application to different arylalkanoic acids is reported in Table 2.4, showing that cyclic aryl ketones can be produced in high yields under very mild conditions.

Cycloacylation through the isocyanate functional group can be utilized for the preparation of partially saturated phenanthridinones that occur in some amaryllidaceae alkaloids, whose synthesis is quite difficult.[28] Cyclization of isocyanates **27** is performed with boron trifluoride etherate and, depending on the structure of the cyclohexene or cyclohexane moieties, furnishes products **28** in 8%–100% yield (Table 2.5). As expected, the efficiency of the process is increased by increasing the number of electron-donating substituents on the aromatic ring. Moreover, the existence and position of a double bond in the cyclohexane unit strongly influence the rate and yield of the reaction, in particular when XY = bond are less, and when YZ = bond are somewhat higher than those observed when no double bonds are present in the starting material. These differences may be due to the energy differences of the cyclohexene ring deformation in the transition states.

Eaton's reagent (a mixture of phosphorous pentoxide and methanesulfonic acid) is frequently utilized as a very efficient solvent catalyst for the preparation of complex polycyclic compounds in the "one-pot" domino process involving a cycloacylation step. Thus, diterpenoid systems,[29] quinolones heterocycles **29** (Scheme 2.15),[30] flavones and chromones,[31] and benzo[*b*] tellurin-4-ones[32] can be elegantly synthesized by exploiting this reagent. However, the process represents a small-scale production method due to

Table 2.4 Cycloacylation promoted by triphenyl phosphine
oxide–triflic anhydride mixture

Carboxylic acid	T (°C)	t (min)	Product	Yield (%)
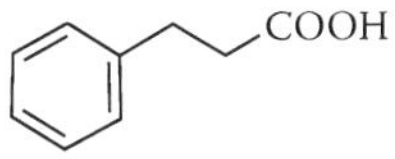	25	240	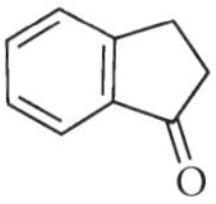	83
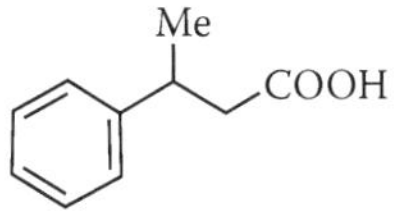	25	240	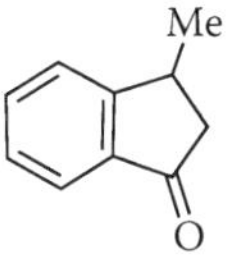	62
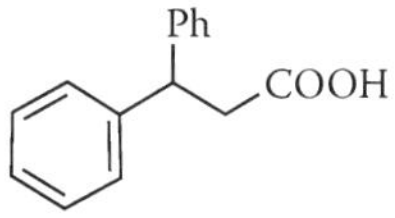	83	30	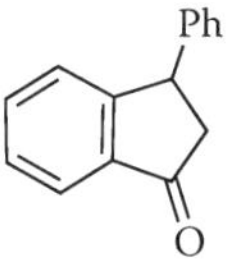	92
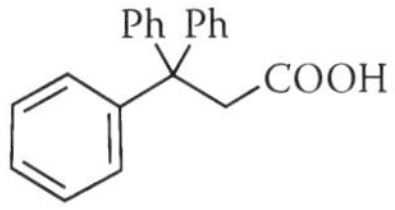	25	30	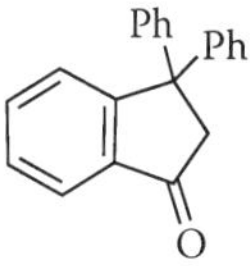	95
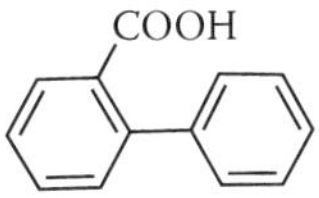	25	30	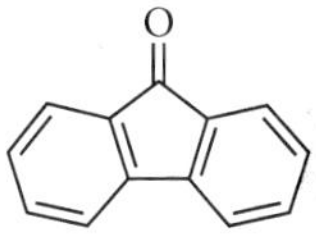	92
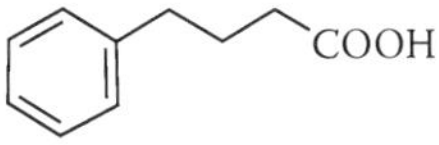	25	240	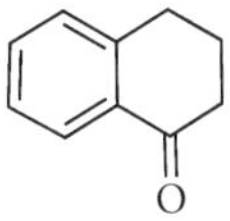	93
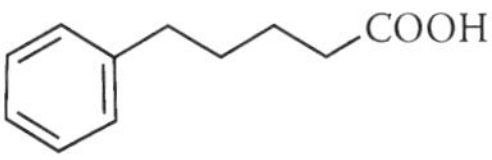	25	240	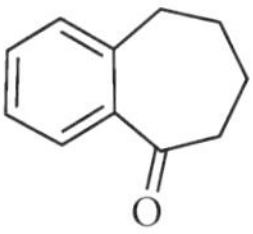	63
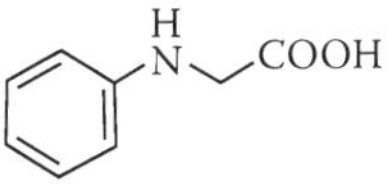	25	2	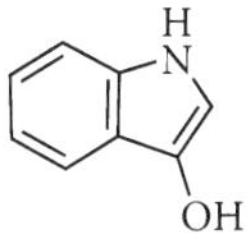	92

Table 2.5 Partially saturated phenanthridinone synthesis
promoted by boron trifluoride etherate

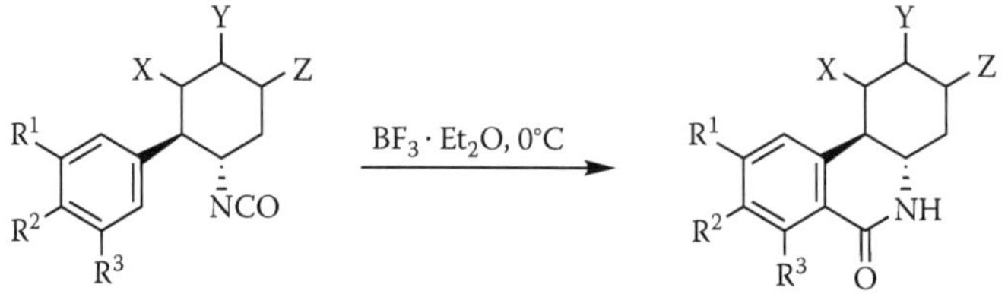

R¹	R²	R³	X	Y	Z	t (h)	Yield (%)
H	H	H	Bond		H	12	8
OMe	OMe	H	Bond		H	8	76
OCH$_2$O		H	Bond		H	8	86
OMe	OMe	OMe	Bond		H	8	52
H	H	H	H	Bond		2	34
OMe	H	H	H	Bond		0.17	64
OMe	OMe	H	H	Bond		0.17	98
OCH$_2$O		H	H	Bond		0.17	97
OMe	H	H	H	H	H	1.5	67
OMe	OMe	H	H	H	H	1.5	100

the great amount of superacid medium needed in the aforementioned efficient processes.

Special attention merits the cycloacylation of 2-benzoylbenzoic acid
derivatives that affords anthraquinones and 1,4-cycloalkanediones. Keto-
acid derivatives such as **30** generally undergo unexpected easy cycloacylation to dicarbonyl compounds **31** (Scheme 2.16). Indeed, compounds **30**
are considered scarcely reactive toward electrophilic aromatic attack,
particularly at the ortho-position to the carbonyl group. However, it
was found that 2-benzoylbenzoyl chloride **33** affords anthraquinone **36**
by aluminum-chloride-promoted cycloacylation via the intermediate
chlorophthalide **35** produced in situ by aluminum-chloride-promoted
"ring-chain tautomerism" **33** ⇄ **35** (Scheme 2.17).[33]

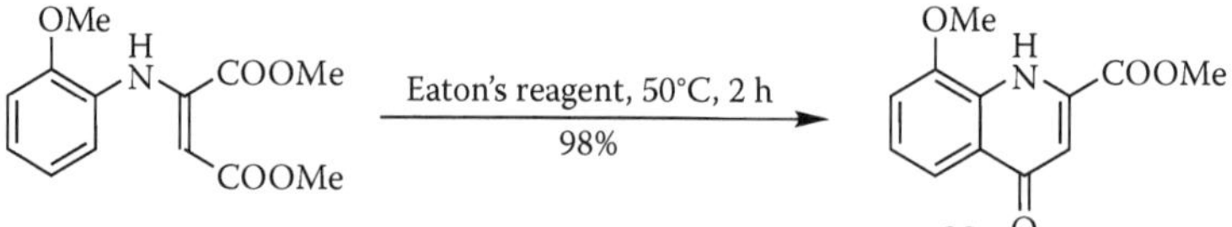

Scheme 2.15

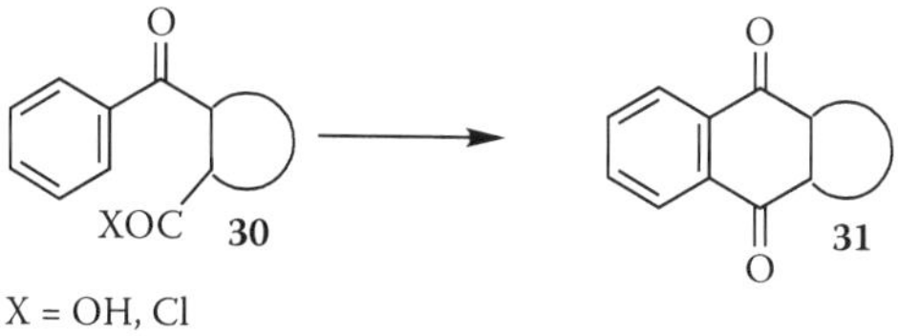

Scheme 2.16

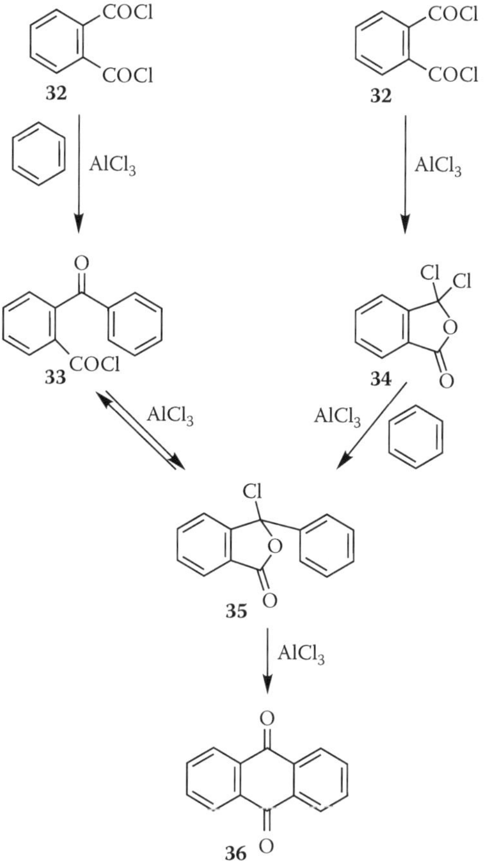

Scheme 2.17

Intermediate **35** can be alternatively produced by aluminum-chloride-promoted reaction of the cyclic dichloride **34** that can be in turn produced by the same ring-chain isomerization of phthalic dichloride **32**. This hypothesis has been strongly supported by ^{13}C NMR studies showing that cyclic dichloride **34** is produced by treatment of phthalic dichloride **32** with

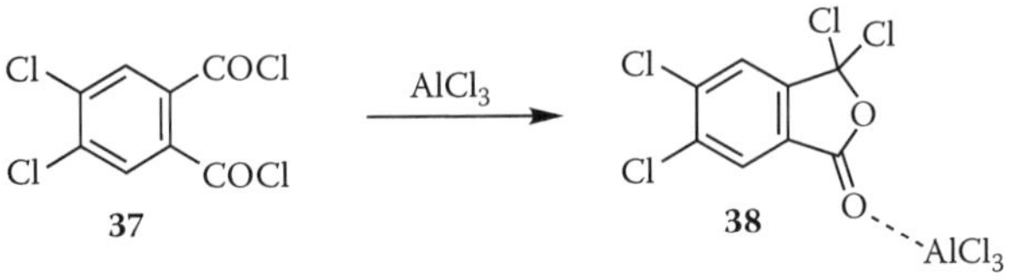

Scheme 2.18

aluminum chloride in deuterated methylene chloride.[34] The production of the active complex **38** by mixing equimolecular amounts of 3,4-dichlorophthalic dichloride **37** and aluminum chloride was further confirmed by multinuclear NMR studies and x-ray analyses[35] (Scheme 2.18).

These studies allow the efficient synthesis of anthraquinones under mild conditions, including anthracyclinones, by direct ortho-bisacylation of aromatic compounds with *ortho*-phthaloyl dichlorides; a few examples are reported in Table 2.6.[34–36] Similarly, aromatic cycloalkanediones can be easily prepared by bis-acylation of aromatic compounds with different aliphatic diacyl chlorides.[37]

In agreement with the phthalide-based mechanism, rufigallol **40**, a compound showing interest in both biological and material science, can be synthesized in 80% yield by self-condensation of gallic acid **39** promoted by concentrated sulfuric acid under 360 W microwave irradiation (Scheme 2.19).[38]

Indane-1,3-diones **44** can be synthesized by a three-component sequential aluminum-chloride-promoted cross condensation–cycloacylation of various acyl chlorides (Scheme 2.20).[39] Thus, the poorly stable acetoacetyl chloride **41**, produced in situ by aluminum chloride-promoted self-condensation of acetyl chloride, reacts at the methylene active carbon with aromatic acyl chlorides **42**, giving the reactive complexes **43** that

Table 2.6 Bis-acylation of different 1,4-dimethoxybenzenes with phthaloyl chloride promoted by aluminum chloride

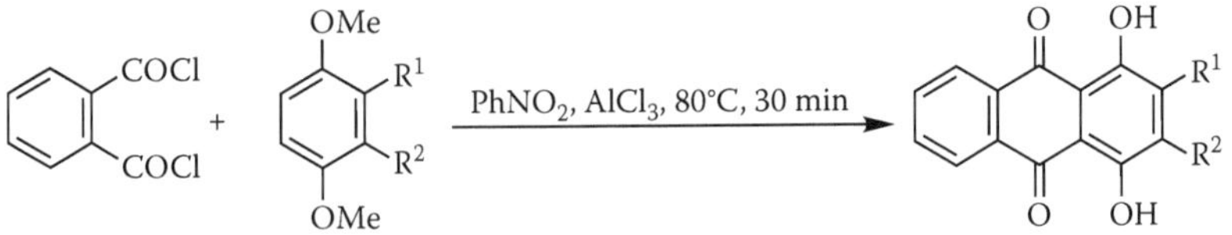

R^1	R^2	Yield (%)	Selectivity (%)
H	H	72	90
Me	H	70	85
(R) $CH_2C(COMe)(OH)CH_2CH_2$		85	97

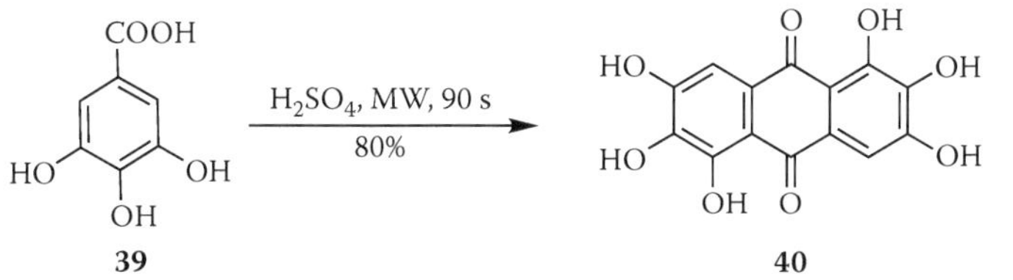

Scheme 2.19

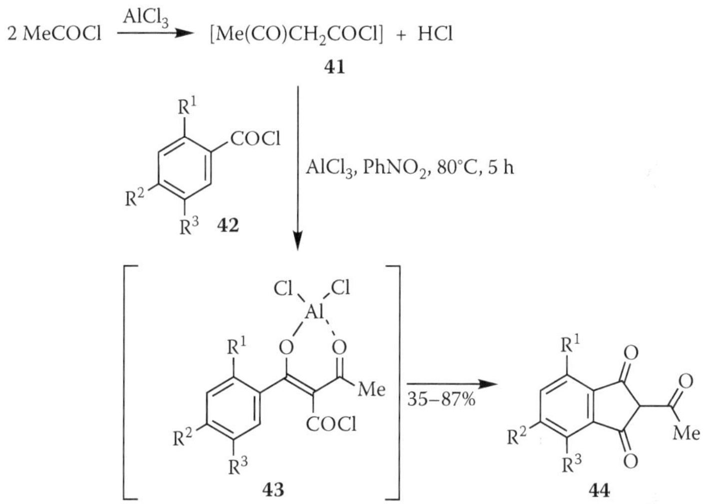

R^1 = H, Me, Cl
R^2 = H, Me, OMe, NO$_2$, Cl
R^3 = H, Cl

Scheme 2.20

undergo easy ortho-cycloacylation, affording 2-acetyl-1,3-indanediones **44** in good yields.

Similarly, malonyl chloride reacts with benzoyl chlorides **45** under aluminum chloride catalysis, giving the selective C-benzoylation at the active methylene group, followed by easy intramolecular electrophilic acylation, affording compounds **47** due to the favorable steric and electronic requisites of the complexes **46** (Scheme 2.21). Cyclic tricarbonyl compounds **47** undergo dechlorocarbonylation during acid quenching to afford indane-1,3-diones **48** in good yields.

In a similar approach, hydroxynaphthoquinones **50**, including heterocyclic analogs, can be easily synthesized by aluminum-chloride-catalyzed double acylation of aromatic and heteroaromatic β-ketoesters **49** with oxalyl chloride (Scheme 2.22).[40]

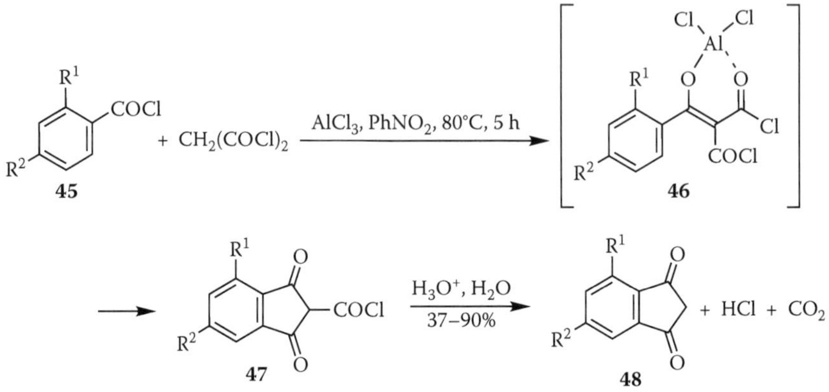

R^1 = H, Cl
R^2 = H, Me, OMe, Cl, NO_2

Scheme 2.21

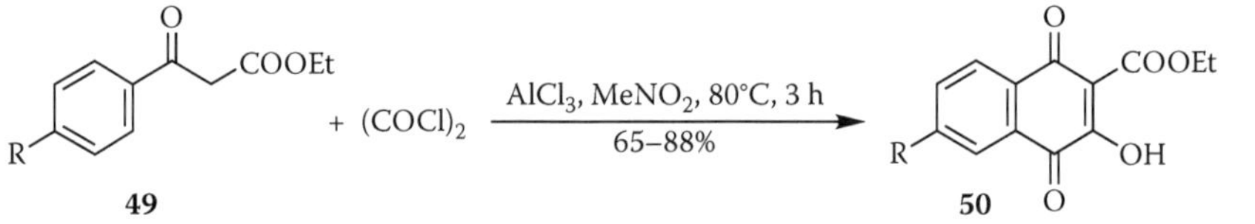

R = H, Me, OMe, Cl

Scheme 2.22

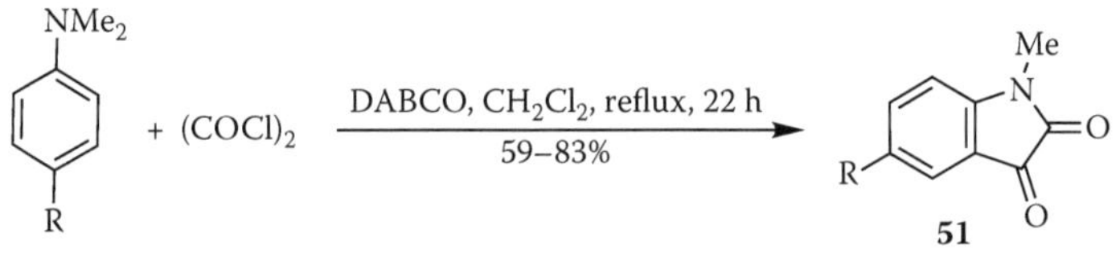

R = H, Me, OMe, F, Cl, Br

Scheme 2.23

A particular case of metal-free bisacylation is reported concerning the reaction of para-substituted *N,N*-dimethylanilines with oxalyl chloride promoted by 1,4-diazabicyclo(2.2.2)octane (DABCO) for the production of compounds **51** (Scheme 2.23).[41] The reaction requires a nearly stoichiometric mixture of oxalyl chloride, DABCO, and *N,N*-dimethylaniline, and proceeds through ortho-C-acylation followed by N-acylation-mono-demethylation. The base presumably takes part in the initial production of the oxalyl chloride acylium salt as well as in the demethylation step.

A quite interesting methodology involves the use of lithium perchlorate anhydrides complexes in the acylation of activated aromatic compounds.[42] Lithium perchlorate is frequently utilized as promoter to accelerate the acylation process and to increase the yield in the reaction catalyzed by metal triflates.[43,44] However, lithium perchlorate itself can act as a very efficient catalyst in the acylation of variously substituted methoxy- and methylbenzenes with AAN affording the aryl ketones in 65%–99% yield. The exceptional activation is ascribed to the formation of a complex with a strong electrophilic character between lithium perchlorate and AAN in neat AAN. The observed para-regioselectivity can be interpreted in terms of the high steric requirement of the lithium perchlorate–AAN complex. A further important feature of this process is the possibility of quantitatively recovering and reusing the catalyst after activation. It must be underlined, however, that lithium perchlorate is an oxidizing compound and can undergo devastating explosions; consequently, it must be handled with maximum care.

A number of studies were published showing the multistep and multicomponent preparation of complex target compounds where Friedel–Crafts acylation plays the driving role. As an example, the cascade approach to the phenanthrene skeleton can be achieved by exploiting a tandem acylation–cycloalkylation process between cyclohexene-1-acetic acid **52** and aromatic substrates promoted by polyphosphoric acid (Table 2.7).[45] The reaction requires a stoichiometric amount of polyphosphoric acid to activate the carboxylic acid **52** toward electrophilic acylation. The intermediate ketones **53** undergo subsequent cycloalkylation, affording 1,2,3,4,4a,10a-hexahydrophenanthrene-9(10H)-ones **54** as a mixture of cis and trans isomers. Compounds **54** are valuable intermediates for the synthesis of morphines, diterpenes, perhydrophenanthrenes, and d-homosteroids.

The preparation of 5-chloroindanone **55**, useful in the production of agrochemicals, can be simply performed in 75% yield by electrophilic cyclo(acylation–alkylation) of chlorobenzene with 3-chloropropanoyl chloride in the presence of an excess of hydrofluoric acid and a stoichiometric amount of boron trifluoride without solvent (Scheme 2.24).[46]

Table 2.7 Hexahydrophenanthrenones preparation through polyphosphoric acid via acylation–cycloalkylation of aromatics with cyclohexene-1-acetic acid

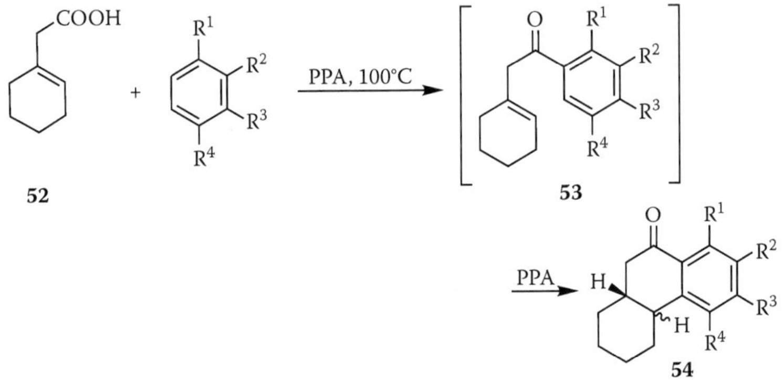

R¹	R²	R³	R⁴	t (h)	Yield (%)
H	H	H	H	8	50
H	H	Me	H	4	62
H	Me	Me	H	3	65
H	H	Cl	H	4	55
Cl	H	H	Cl	4	52
H	H	Br	H	3	58
H	H	NMe₂	H	3	62
H	H	NEt₂	H	3	64
H	H	OMe	H	2.5	64
OMe	H	H	OMe	2	64
OMe	H	OMe	H	2	62

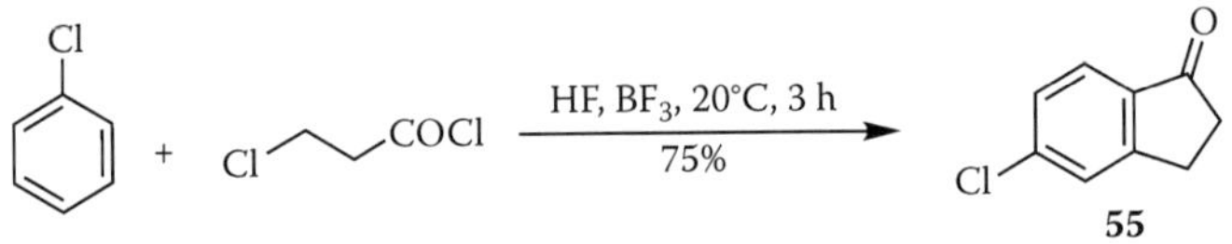

Scheme 2.24

References

1. Jefford, C. W., Tang, Q., and Boukouvalas, J. 1990. Regioselective 2-acylation of N-substituted pyrroles by intramolecular delivery. *Tetrahedron Lett.* 31: 995–998.
2. Josey, A. D. 1973. 1-(2-Methoxycarbonylphenyl)pyrrole. In *Organic syntheses*, Collective volume 5, ed. H. E. Baumgarten, 716–717. New York: John Wiley & Sons.

3. Nakatsuka, S.-I., Teranishi, K., and Goto, T. 1994. Formation of 6-acylindoles from 1-acylindoles. *Tetrahedron Lett.* 35: 2699–2700.

4. Okauchi, T., Itonaga, M., Minami, T., Owa, T., Kitoh, K., and Yoshino, H. 2000. A general method for acylation of indoles at the 3-position with acyl chlorides in the presence of dialkylaluminum chloride. *Org. Lett.* 2: 1485–1487.

5. Yeung, K.-S., Farkas, M. E., Qiu, Z., and Yang, Z. 2002. Friedel–Crafts acylation of indoles in acidic imidazolium chloroaluminate ionic liquid at room temperature. *Tetrahedron Lett.* 43: 5793–5795.

6. Mohile, S. S., Potdar, M. K., and Salunkhe, M. M. 2003. An alternative route to syntheses of aryl keto acids in a chloroaluminate ionic liquid. *J. Chem. Res. (S)* 650–651.

7. Surette, J. K. D., Green, L., and Singer, R. D. 1996. 1-Ethyl-3-methylimidazolium hologenoaluminate melts as reaction media for the Friedel–Crafts acylation of ferrocene. *Chem. Commun.* 2753–2754.

8. Akhrem, I., Orlinkov, A., and Vol'pin, M. 1993. Direct alkylacylation of arenes and cycloalkanes in the presence of aprotic superacids. *J. Chem. Soc., Chem. Commun.* 257–258.

9. Galli, C. 1979. Acylation of arenes and heteroarenes with in situ generated acyl trifluoroacetates. *Synthesis* 303–304.

10. Tedder, J. M. 1955. The use of trifluoroacetic anhydride and related compounds in organic syntheses. *Chem. Rev.* 55: 787–827.

11. Smyth, T. P. and Corby, B. W. 1998. Toward a clean alternative to Friedel–Crafts acylation: in situ formation, observation, and reaction of an acyl bis(trifluoroacetyl)phosphate and related structures. *J. Org. Chem.* 63: 8946–8951.

12. Veeramaneni, V. R., Pal, M., and Yeleswarapu, K. R. 2003. A high speed parallel synthesis of 1,2-diaryl-1-ethanones via a clean-chemistry C-C bond formation reaction. *Tetrahedron* 59: 3283–3290.

13. Goldfinger, M. B. 2000. Manufacture of polyketones using carboxylic acid anhydride/phosphoric acid catalysts. WO Patent 52,075.

14. Hwang, J. P., Prakash, G. K. S., and Olah, G. A. 2000. Trifluoromethanesulfonic acid catalyzed novel Friedel–Crafts acylation of aromatics with methyl benzoate. *Tetrahedron* 56: 7199–7203.

15. Anderson, K. W. and Tepe, J. J. 2002. The first intermolecular Friedel–Crafts acylation with β-lactams. *Org. Lett.* 4: 459–461.

16. Gors, H. C., Horner, P. J., and Jansons, V. 1991. Friedel–Crafts preparation of aromatic ketones with an inorganic salt controlling agent. U.S. Patent 5,068,447.

17. Lindley C. R. 1992. BF_3 catalyzed acylation of aromatic compounds. EP Patent 488,638.

18. Marshall, W. S., Sigmund, S. K., and Whitesitt, C. A. 1988. Process for leukotriene antagonists. U.S. Patent 4,777,299.

19. Marshall, W. S., Sigmund, S. K., and Whitesitt, C. A. 1988. Process for intermediates to leukotriene antagonists. U.S. Patent 4,777,298.

20. Gissot, A., Becht, J.-M., Desmurs, J. R., Pévère, V., Wagner, A., and Mioskowski, C. 2002. Directed ortho-metalation, a new insight into organosodium chemistry. *Angew. Chem. Int. Ed.* 41: 340–343.

21. Suzuki, H., Iwata, C., Sakurai, K., Tokumoto, K., Takahashi, H., Hanada, M., Yokoyama, Y., and Murakami, Y. 1997. A general synthetic route for l-substituted 4-oxygenated β-carbolines (Synthetic studies on indoles and related compounds 41). *Tetrahedron* 53: 1593–1606.
22. Olah, G. A. 1964. *Friedel-Crafts and related reactions*. London: Interscience, 3: 911–960.
23. Fillion, E. and Fishlock, D. 2003. Convenient access to polysubstituted 1-indanones by Sc(OTf)₃-catalyzed intramolecular Friedel–Crafts acylation of benzyl Meldrum's acid derivatives. *Org. Lett.* 5: 4653–4656.
24. Sharma, A. K., Subramani, A. V., and Gorman, C. B. 2007. Efficient synthesis of halo indanones via chlorosulfonic acid mediated Friedel–Crafts cyclization of aryl propionic acids and their use in alkylation reactions. *Tetrahedron* 63: 389–395.
25. Molloy, B. B. 1979. Trifluoromethyl substituted 1-aminoindanes. U.S. Patent 4,132,737.
26. Hendrickson, J. B. and Hussoin, M. S. 1989. Reactions of carboxylic acids with "phosphonium anhydrides." *J. Org. Chem.* 54: 1144–1149.
27. Uhlig, F. and Snyder, H. R. 1960. Polyphosphoric acid as a reagent in organic chemistry. In *Advances in organic chemical methods and results*, ed. R. A. Raphael, E. C. Taylor, and H. Wynberg, 35–81. London: Interscience.
28. Balázs, L., Nyerges, M., Kádas, I., and Töke, L. 1995. Synthesis of phenanthridin-6(5H)-ones. *Synthesis* 1373–1375.
29. Bhar, S. S. and Ramana, M. M. V. 2004. Novel domino reactions for diterpene Synthesis. *J. Org. Chem.* 69: 8935–8937.
30. Zewge, D., Chen, C.-Y., Deer, C., Dormer, P. G., and Hughes, D. L. 2007. A mild and efficient synthesis of 4-quinolones and quinolone heterocycles. *J. Org. Chem.* 72: 4276–4279.
31. McGarry, L. W. and Detty, M. R. 1990. Synthesis of highly functionalized flavones and chromones using cycloacylation reactions and C-3 functionalization. A total synthesis of hormothamnione. *J. Org. Chem.* 55: 4349–4358.
32. Shtern, D., Manchala, G., and Detty, M. R. 1998. Synthetic routes to 4H-7-hydroxybenzo[b]tellurin-4-ones. *Organometallics* 17: 3588–3592.
33. Valters, R. E. and Flitsch, W. 1985. *Ring-chain tautomerism*, ed. Katritzky, A. R. New York: Plenum Press.
34. Sartori, G., Casnati, G., Bigi, F., and Foglio, F. 1990. A new methodological approach to anthraquinone and antrhacyclinone synthesis. *Gazz. Chim. Ital.* 120: 13–19.
35. Sartori, G., Bigi, F., Tao, X., Porta, C., Maggi, R., Predieri, G., Lanfranchi, M., and Pellinghelli, M. A. 1995. An investigation of the reaction mechanism of the bis-acylation of aromatics with o-phthaloyl dichlorides: regioselective synthesis of anthraquinones. *J. Org. Chem.* 60: 6588–6591.
36. Sartori, G., Casnati, G., Bigi, F., and Robles, P. 1987. Metal template ortho-acylation of phenols; a new general approach to anthracyclinones. *Tetrahedron Lett.* 28: 1533–1536.
37. Sartori, G., Bigi, F., Goffredi, G., Maggi, R., Portioli, R., and Casnati, G. 1993. A direct route for the regioselective synthesis of hydroxylated aromatic cycloalkanediones. *J. Chem. Res. (S)* 324.
38. Bisoyi, H. K. and Kumar, S. 2007. Microwave-assisted synthesis of rufigallol and its novel room-temperature liquid crystalline derivatives. *Tetrahedron Lett.* 48: 4399–4402.

39. Sartori, G., Bigi, F., Maggi, R., Baraldi, D., and Casnati, G. 1992. Acylation of aroyl chlorides via a template Friedel–Crafts process: synthesis of indan-1,3-diones. *J. Chem. Soc., Perkin Trans. 1* 2985–2988.

40. Sartori, G., Bigi, F., Canali, G., Maggi, R., Casnati, G., and Tao, X. 1993. Friedel–Crafts coordinated process: highly selective synthesis of hydroxynaphthoquinones. *J. Org. Chem.* 58: 840–843.

41. Cheng, Y., Ye, H.-L., Zhan, Y.-H., and Meth-Cohn, O. 2001. A very simple route to *N*-methylisatins: Friedel–Crafts acylation of *p*-substituted *N,N*-dimethylanilines with oxalyl chloride and DABCO. *Synthesis* 904–908.

42. Bartoli, G., Bosco, M., Marcantoni, E., Massaccesi, M., Rinaldi, S., and Sambri, L. 2002. LiClO$_4$-acyl anhydrides complexes as powerful acylating reagents of aromatic compounds in solvent free conditions. *Tetrahedron Lett.* 43: 6331–6333.

43. Chapman, C. J., Frost, C. G., Hartley, J. P., and Whittle, A. J. 2001. Efficient aromatic and heteroatom acylations using catalytic indium complexes with lithium perchlorate. *Tetrahedron Lett.* 42: 773–775.

44. Matsuo, J.-I., Odashima, K., and Kobayashi, S. 2000. Gallium nonafluorobutanesulfonate as an efficient catalyst in Friedel–Crafts acylation. *Synlett* 403–405.

45. Ramana, M. N. V. and Potnis, P. V. 1996. Tandem acylation–cycloalkylation: a novel synthesis of phenanthrenes. *Synthesis* 1090–1092.

46. Laurain, N. and Saint-Jalmes, L. 2000. Procedé de preparation d'un composé de type indanone ou thioindanone.WO Patent 43,342.

chapter 3

Catalytic homogeneous acylations

As already underlined in the introduction to this book, catalytic homogeneous acylation reactions represent a remarkable improvement in the preparation of aromatic ketones because, in the conventional Lewis-acid-promoted reactions, formation of a stable complex between the ketone product and the catalyst implies that at least a stoichiometric amount of catalyst must be utilized. This drawback prompted a great number of studies aimed at setting up the experimental conditions to make catalytic Friedel–Crafts acylation reactions. Some positive results from the homogeneous catalytic Friedel–Crafts acylations are described here, with special attention to crucial economic and environmental advantages such as the recycling of expensive catalysts and the development of solvent-free and highly selective synthetic processes.

A good and comprehensive review of catalytic electrophilic acylation was published by Pearson and Buehler.[1] Only the catalysts most widely used were considered, with special attention to iron trichloride, zinc chloride, iodine, and elemental iron. The substrates that can be acylated using small amounts of catalysts include alkylarenes, aryl ethers, biphenyls, naphthalenes, acenaphthenes, fluorene, furans, and thiophenes. Aromatic acyl chlorides lead to better yields than aliphatic ones, reaching a maximum of 96% and a minimum of 34%. In general, the reactions are performed at relatively high temperatures (from 50°C to ~200°C) at which hydrogen chloride evolution is fairly rapid.

Concerning catalytic activity, it appears that the relatively high reaction temperature favors dissociation of the ketone–Lewis acid complex, which permits Lewis acid to be recycled. A similar explanation can account for iodine recycling.

3.1 Metal halides

Iron(III) salts represent efficient Lewis acid catalysts frequently utilized in a very low amount to promote Friedel–Crafts acylation. Thus, (*S*)-2,4,6-trimethylphenyl-1-phthalimidoethyl ketone **3** has been synthesized in 79% yield by aminoacylation of mesitylene **1** with chiral α-phthalimidocarbonyl chloride **2** in the presence of iron trichloride (1% mol) at 0°C for 48 h (Scheme 3.1).[2]

The same catalyst (10% mol) can be utilized in the benzoylation of anisole by benzoyl chloride (BC), showing yield improvement when performed in combination with microwave irradiation.[3,4] The BC conversion is 95% (with ortho/para isomer ratio 0.06) after 1 min irradiation at 165°C and 300 W power. The reaction can be applied to different activated arenes, giving products in high yields and showing the characteristics of an economical energy-saving process able to perform clean reactions in reduced time. A further advantage is the avoidance of the solvent since the aromatic substrate can replace it.

The benzoylation of 2-methoxynaphthalene (2-MN) **4** with an equimolecular amount of BC in the presence of aluminum chloride (5% mol) gives products **5, 6,** and **7** in 63:25:12 ratio (Table 3.1).[5] With indium trichloride in place of aluminum chloride, compound **5** is obtained in a higher selectivity (88%). On increasing the temperature in the reaction with indium trichloride, the product composition is changed, and compound **6** is the predominant one (84% selectivity). Quite similar results are achieved with iron trichloride, tin tetrachloride, and zinc chloride, while in the case of antimony pentachloride and titanium tetrachloride, compound **5** is the major product. Mechanistic studies show that the reaction involves the isomerization of **5** to **7** and **6**.

Toluene as well as xylenes, mesitylene, naphthalene, and anisole are efficiently acylated in the presence of bismuth trichloride (10%–20% mol), a medium-strength Lewis acid.[6] Moreover, bismuth trichloride being

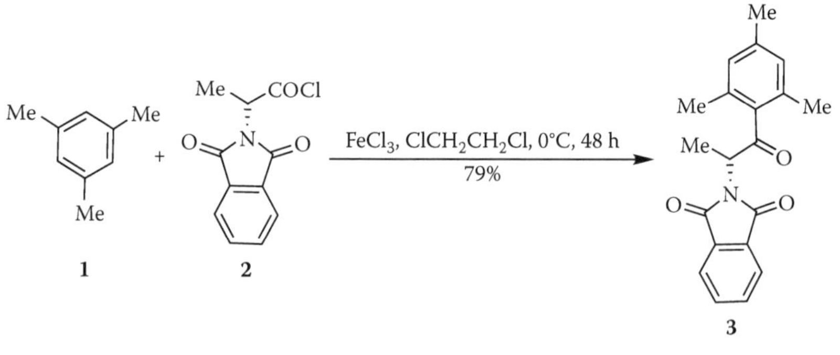

Scheme 3.1

Table 3.1 Activity and selectivity of different Lewis acids in the benzoylation of 2-MN with BC

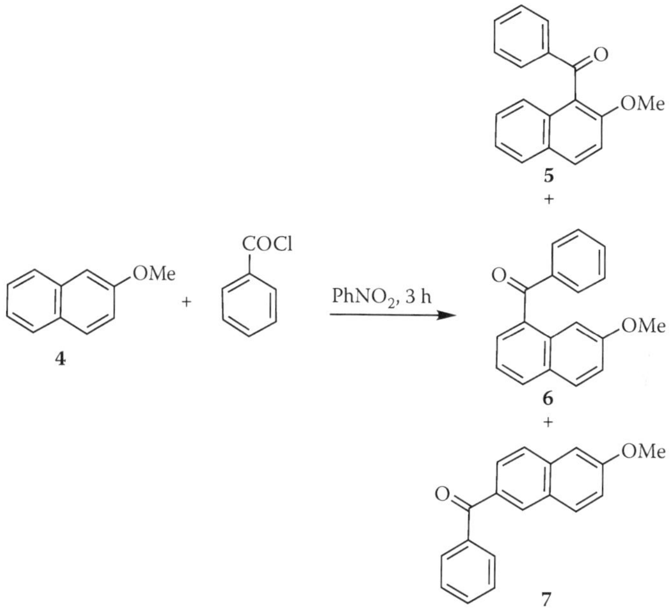

				Product distribution (%)		
Catalyst	T (°C)	4 Conv. (%)	Yield (%)	**5**	**6**	**7**
AlCl$_3$	50	56	91	63	25	12
InCl$_3$	50	58	99	88	9	3
InCl$_3$	160	72	79	—	84	16
SbCl$_5$	160	56	89	70	22	8
TiCl$_4$	160	53	83	86	9	5
FeCl$_3$	160	66	82	—	80	20
SnCl$_4$	160	70	80	4	77	19
ZnCl$_2$	160	73	84	1	85	14

freely soluble in a hydrogen-chloride-containing solution, the recovery of bismuth from the reaction mixture is possible simply by neutralization of the aqueous layer with sodium hydroxide, which allows for quantitative precipitation of bismuth(III) oxychloride, a water-insensitive and ecofriendly compound.[7] Interestingly, this compound can be transformed in situ into bismuth trichloride by reaction with acyl chlorides: BiOCl + 2RCOCl → BiCl$_3$ + (RCO)$_2$O. This transformation can also be performed by using thionyl chloride as a chlorinating agent,[8] which prevents loss of any of the acyl chloride, particularly rare and costly types, in the successive acylation reaction.

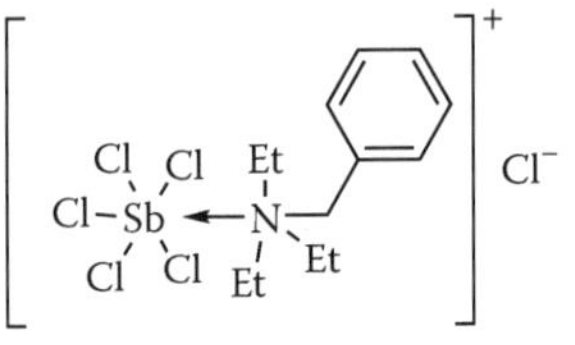

Figure 3.1 Antimony pentachloride–benzyltriethyl ammonium chloride complex.

The antimony pentachloride–benzyltriethyl ammonium chloride complex (SbCl$_5$–TEBA, Figure 3.1), shows quite interesting catalytic efficiency in the acylation of activated aromatic compounds (i.e., toluene, xylenes, aryl ethers) with aromatic and chloroacetyl chlorides.[9] Reactions are carried out with SbCl$_5$–TEBA (5% mol) in boiling nitromethane, giving ketones in 73%–96% yield. The catalyst has many advantages, such as ready access, minimal toxicity, reusability, insensitivity to atmosphere and moisture.

A further antimony derivative, namely, bis-(*para*-methoxyphenyl) boryl hexachloroantimonate [prepared in situ from antimony pentachloride and bis-(*para*-methoxyphenyl)boryl chloride] can be utilized (25% mol) in the acylation of anisole and veratrole with acetone acyl enolates in 52%–88% yield.[10] The major advantage of the method resides in the possibility of performing Friedel–Crafts acylation at room temperature in a reaction medium that can be kept almost neutral throughout the reaction, acetone being the only co-product.

Hard Lewis acid chloroaluminate ionic liquids show intense catalytic activity in the Friedel–Crafts acylation reaction; however, they suffer from the same issues as anhydrous aluminum chloride.[11] Of particular interest to these reactions, aluminum chloride may be replaced by indium trichloride to form chloroindate(III) ionic liquids.[12] The advantage of using indium trichloride compared with aluminum chloride is represented by its hydrolytic stability and reduced oxophilicity. Chloroindate(III) ionic liquids are synthesized by mixing 1-butyl-3-methylimidazolium chloride {[C$_4$mim]Cl} with anhydrous indium trichloride at 80°C.[13] In the benzoylation of anisole with benzoic anhydride (BAN) at 80°C, the best yield of 4-methoxybenzophenone is achieved when indium trichloride (5% mol) is used in conjunction with ionic liquids either as a solution in [C$_4$mim][NTf$_2$] or as chloroindate(III) binary ionic liquid for 3 h (~80% yield; ortho/para = 0.06). The chloroindate(III) catalyst can be recycled five times with only a loss in activity from 79% to 62% yield, and no change in selectivity. Benzene, isobutylbenzene, toluene, and naphthalene all give good yields with BC (81%–96%), whereas with BAN, lower yields are obtained (22%–87%). Worth noting is the fact that naphthalene gives the less favored 2-substituted product in the reaction with the more bulky anhydride in respect to the chloride.[11]

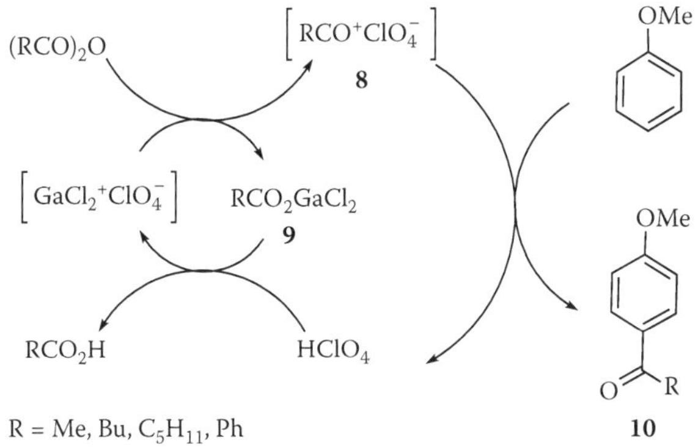

Scheme 3.2

Catalytic Friedel–Crafts acylation can be greatly improved by using a combination of Lewis acid with silver or lithium perchlorates. Activated aromatic substrates (aromatic ethers and alkyl arenes) give acylation by reaction with aliphatic and aromatic anhydrides or chlorides in the presence of a combination of gallium trichloride (10% mol) and silver perchlorate (10% mol), affording aryl ketones in 80%–100% yield at room temperature.[14] A possible rationale of the process with anisole as model reagent is depicted in Scheme 3.2. The active catalyst, dichlorogallium(III) perchlorate ($GaCl_2{}^+ClO_4{}^-$), initially reacts with the anhydride to give the intermediate **8**. Then **8** in turn reacts with the aromatic compound to afford the desired ketone **10** along with perchloric acid, which reacts with the acylgallium(III) salt **9,** regenerating the catalyst. Similar good results can be achieved by a combination of antimony pentachloride (4% mol) and lithium perchlorate (100% mol) in refluxing methylene chloride.[15] An interesting fact is that, in the present reaction, anhydrides give better yields than the corresponding acyl chlorides. For example, anisole reacts with valeric anhydride to give *para*-methoxyphenyl butyl ketone in 85% yield, while 11% yield of the same product is obtained in the reaction with valeroyl chloride under the same reaction conditions.

Higher catalytic activity (91% yield) is achieved by using a mixture of niobium pentachloride (1% mol) and silver perchlorate (3% mol) in nitromethane at 80°C.[16]

Late-transition metal salts have been utilized as catalysts to promote Friedel–Crafts acylation of arenes and heteroarenes with anhydrides.[17] A mismatch between their soft metal center and the hard carbonyl oxygen atoms of the products avoids the formation of a kinetically inert complex and results in catalytic turnovers. Although late-transition metal salts exhibit, a priori, rather poor Lewis acidity, sufficient reactivity can be gained by rendering them cationic.[18] The acylation of variously substituted

Table 3.2 Acylation of aromatics with anhydrides catalyzed by
dichlorobis(benzonitrile)platinum(II)–silver hexafluoroantimonate mixture

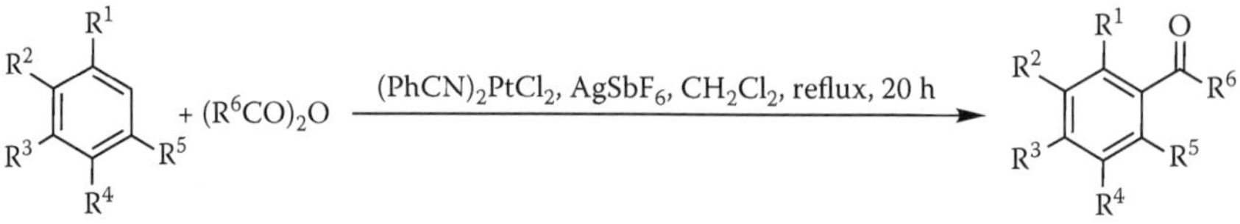

R¹	R²	R³	R⁴	R⁵	R⁶	Yield (%)
Me	H	Me	H	Me	Me	76
Me	Me	Me	Me	Me	Me	70
H	H	OMe	H	H	Me	75
H	H	OMe	H	H	Ph	77
H	H	OMe	H	H	C_5H_{11}	77
OMe	H	OMe	H	OMe	Me	70
OMe	OMe	OMe	H	H	Me	62
H	OCH₂O		H	H	Me	61
H	CH₂COMe	OMe	H	H	Me	61
Me	Me	Me	Me	Me	Me	70

aromatic substrates with anhydrides is thus performed in refluxing methylene chloride and in the presence of dichlorobis(benzonitrile)platinum(II) (2.5% mol) and silver hexafluoroantimonate (5% mol). Representative examples are reported in Table 3.2. The catalytic system works well with moderately or strongly activated arenes. Even the sterically hindered position in pentamethylbezene is smoothly acylated. Also, heteroaromatic compounds such as furan and thiophene are equally well suited, whereas *N,N*-dimethylaniline and indole are inert, most likely because the cationic platinum(II) species are inhibited by coordination with the nitrogen atom of the substrates. Spectroscopic studies in the model reaction with anisole show that these results are best interpreted in terms of a coordination via the π-system rather than the ether oxygen, although the precise hapticity of binding cannot yet be unambiguously deduced.

3.2   *Trifluoroacetic acid, triflic acid, and (metal) derivatives*

Trifluoroacetic and triflic anhydrides can be utilized to promote acylation of aromatics with carboxylic acids without any metal catalyst. Thus, the trifluoroacetic anhydride (TFAA)-promoted cycloacylation of conveniently functionalized carboxylic acids prepared from Baylis–Hillman adducts

Table 3.3 Synthesis of
(E)-3-benzylidenechroman-4-ones promoted by TFAA

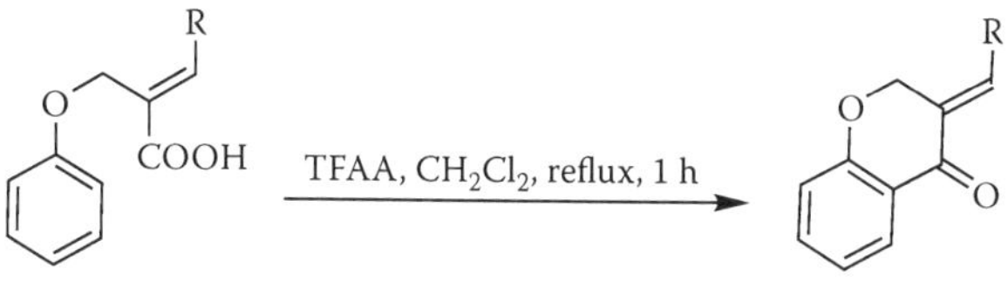

R	Yield (%)
Ph	91
4-MeC$_6$H$_4$	94
2-MeC$_6$H$_4$	90
4-EtC$_6$H$_4$	93
4-Pr^{i}C$_6$H$_4$	91
4-OMeC$_6$H$_4$	92

represents the crucial step in the preparation of classes of biologically active compounds.[19,20]

A representative class of (E)-3-benzylidenechroman-4-ones **12** can be prepared with 90%–94% yield, starting from (2E)-3-aryl-2-phenoxymethyl-prop-2-enoic acids **11** (Table 3.3).[19] The reaction is simply performed by refluxing compounds **11** for 1 h in methylene chloride in the presence of TFAA.

In a similar way, a series of (E)-2-benzylideneindan-1-ones **14** can be prepared in 50%–67% yield from Baylis–Hillman adducts **13** in the presence of TFAA (Table 3.4).[20]

Acylation of electron-rich aromatics is reported to occur efficiently and very easily through mixed carboxylic-triflic anhydrides without any catalyst.[21] The reaction can be applied to methoxy- and alkylarenes and thiophene, with acetic and benzoic acid, in a neat mixture, or in nitromethane (65%–98% yield) at room temperature or at 45°C.

Metal triflates can be easily prepared from metal halides and triflic acid at −78°C.[22] They show several unique properties compared with the corresponding metal halides. In an early study, Olah reported the use of boron-, aluminum-, and gallium triflates [M(OTf)$_3$] as effective Friedel–Crafts catalysts.[22] In the benzoylation and acetylation of toluene and benzene with acyl chlorides, the relative reactivity is boron triflate > gallium triflate > aluminum triflate, in agreement with the relative acidity strength.

Use of bismuth(III) triflate allows achieved acylation of both activated and deactivated aromatic compounds with anhydrides and acyl chlorides.[23] Thus, the acylation of aromatics such as trifluoromethoxybenzene, toluene, benzene, fluorobenzene, and chlorobenzene can be achieved in high

Table 3.4 Synthesis of (E)-2-benzylideneindan-1-ones promoted by TFAA

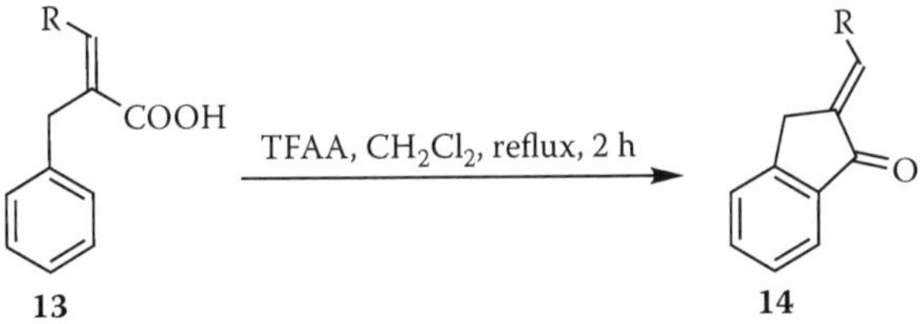

R COOH

$\xrightarrow{\text{TFAA, CH}_2\text{Cl}_2\text{, reflux, 2 h}}$

13

R

O

14

R	Yield (%)
Ph	63
4-MeC$_6$H$_4$	61
4-EtC$_6$H$_4$	59
4-Pr^{i}C$_6$H$_4$	56
2-MeC$_6$H$_4$	67
4-BrC$_6$H$_4$	50

yields (78%–95%) with BAN or BC in the presence of bismuth(III) triflate (10% mol) without solvent. The para-acylation product is the most abundant in all cases (i.e., ortho/para = 0.06 for trifluoromethoxybenzene, 0.01 for fluorobenzene, and 0.15 for chlorobenzene).

Titanium(IV) monochlorotriflate [TiCl(OTf)$_3$] (1% mol), in combination with triflic acid [TfOH] (10% mol), is highly efficient in the acylation of anisole with hexanoic anhydride at room temperature in acetonitrile or 1,2-dichloroethane (98% yield, 100% para-selectivity).[24] It is assumed that the active cationic catalyst of titanium(IV) monochlorotriflate activates the carboxylic anhydride and generates a catalytic amount of mixed triflic–carboxylic anhydride along with titanium carboxylate. The active mixed anhydride immediately reacts with the aromatic substrate, giving the aromatic ketone and regenerating the catalyst by triflic acid. Several examples of this Friedel–Crafts acylation reaction are shown in Table 3.5. When toluene is used as solvent-reagent, the reaction with hexanoic anhydride gives 1-(4-methylphenyl)hexane-1-one in 66% yield.

A dramatic acceleration of benzene acylation is also observed by using a similar catalyst consisting of a combination of hafnium triflate and triflic acid.[25] While in the benzoylation of benzene with BC, low yields of benzophenone (BP) (5%–10%) are obtained in the presence of hafnium triflate (5% mol) or triflic acid (5% mol), a 77% yield is obtained when combining hafnium triflate (5% mol) with triflic acid (5% mol). The yield is improved to 82% when hafnium triflate and triflic acid are used in higher quantity (10% mol); a further increase of the catalyst amount does not improve the

Table 3.5 Acylation of aromatic ethers with anhydrides in the presence of titanium(IV) monochlorotriflate–triflic acid mixture

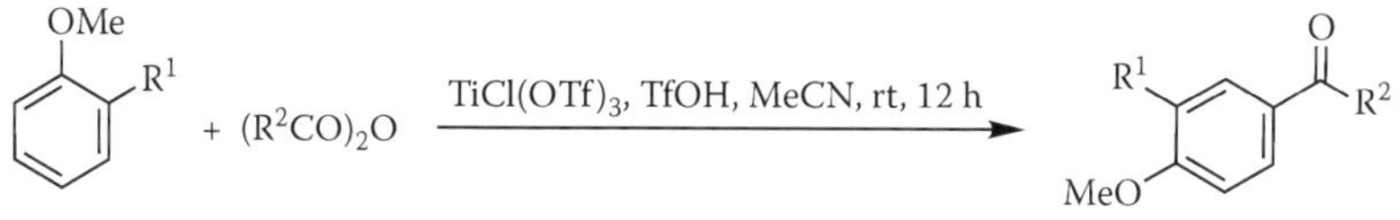

R^1	R^2	Yield (%)
H	C_5H_{11}	98
H	Pr^i	90
H	Bu^t	68
H	Ph	90
OMe	C_5H_{11}	94
OMe	Pr^i	89
OMe	Bu^t	66
OMe	Ph	94
Me	C_5H_{11}	97
Me	Pr^i	96
Me	Bu^t	64
Me	Ph	88

yield. Unactivated aromatic substrates such as chloro- and fluorobenzene react smoothly under these conditions with both aliphatic and aromatic acyl chlorides, affording the corresponding arylketones **15** and **16** in high yields (Table 3.6).

The mixed carboxylic-triflic anhydride produced between the acyl chloride and triflic acid has been earlier described to give electrophilic acylation without any catalyst.[26] This acylation method is advantageous in terms of the mild conditions employed and the easy availability of acyl chlorides. The aromatic substrates are mainly limited to electron-rich arenes. However, the methodology can be applied to unactivated aromatics such as benzene and chlorobenzene under special conditions.

The benzoylation of both activated and deactivated aromatic compounds with benzoic acids (BACs) is achieved by using bismuth triflate (10% mol) in the presence of TFAA or heptafluorobutyric anhydride (HFBA).[27] *meta*-Xylene undergoes acetylation with 100% yield by using acetic acid (AAC) in the presence of TFAA under bismuth or scandium triflate catalysis (10% mol). The bismuth-triflate-catalyzed reaction can be extended to different aromatics as well as to aliphatic and aromatic carboxylic acids, giving ketones in nearly quantitative yield. Benzene and chlorobenzene are benzoylated with a BAC/HFBA mixture in the presence of bismuth triflate (10% mol), giving the corresponding BPs in 90%

Table 3.6 Acylation of aromatics with acyl chlorides promoted by hafnium triflate–triflic acid mixture

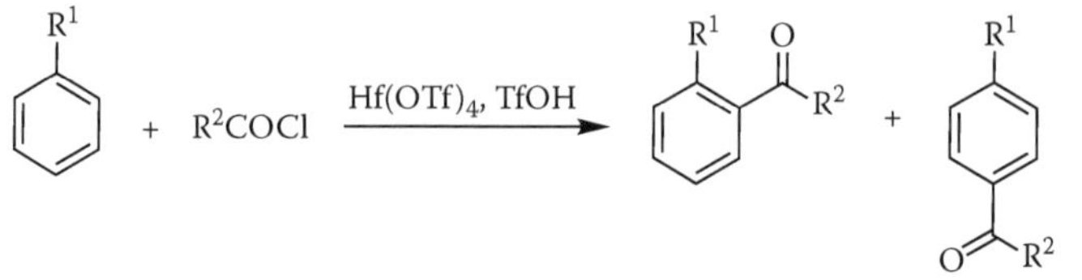

R^1	R^2	T (°C)	t (h)	Yield (%)	Product distribution (%)	
					15	**16**
H	Ph	80	8	82	—	—
H	4-ClC$_6$H$_4$	80	8	78	—	—
Cl	Ph	120	15	72	7	93
Cl	4-ClC$_6$H$_4$	120	15	82	<1	>99
F	Ph	120	15	83	<1	>99
H	C$_5$H$_{11}$	80	8	60	—	—
H	Bu^s	80	6	69	—	—
Me	Ph	100	8	82	15	85
Me	4-ClC$_6$H$_4$	100	8	80	8	92

and 65% yield (ortho/para molar ratio 0.06), respectively. The catalyst can be recovered by precipitation with hexane from the final reaction mixture and reused in the benzoylation of *meta*-xylene, giving the product with unchanged yield (~95%) for four successive cycles.

In the preparation of mixed carboxylic–perfluoroalkane sulfonic anhydrides[28,29] from the corresponding acids, water formation is responsible for the low yields of the acylation product.[30] By removing water from the reaction mixture by azeotropic distillation and adsorption onto molecular sieves, the yield of benzoylation of *para*-xylene with BAC in the presence of perfluorobutanesulfonic acid (10% mol) is improved to 90%.

The direct 1,4-dimethylanthraquinone **17** synthesis via double acylation of *para*-xylene with phthalic anhydride can be performed in 15% yield at 138°C, but an equimolecular amount of triflic acid is necessary (Scheme 3.3). Better results are achieved by using phthaloyl dichloride under triflic acid catalysis (5% mol) at the same temperature: anthraquinone **17** can be synthesized in 52% yield. Moreover, product **17** is obtained in 89% yield when the same reaction is carried out in 1,2-dichloroethane and a higher amount of triflic acid (10% mol).

Lanthanide triflates were developed by Kobayashi as efficient catalysts in Friedel–Crafts acylation.[31] Lanthanide compounds are expected

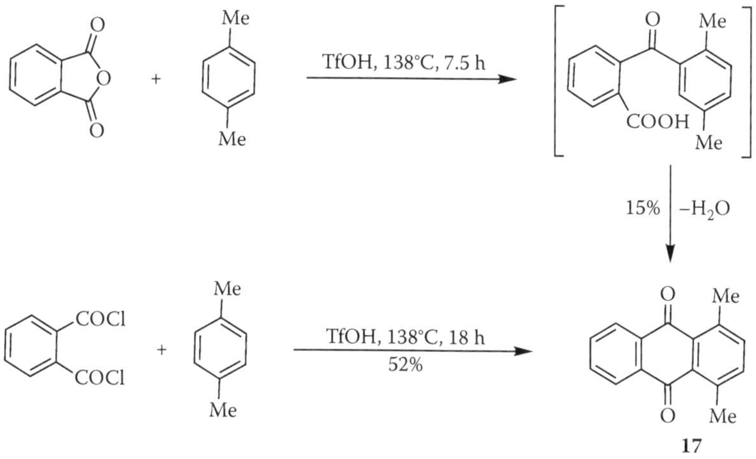

Scheme 3.3

to act as strong Lewis acids because of their hard character and strong affinity with carbonyl oxygen.[32] Moreover, their hydrolysis is slow due to their high hydration energies and hydrolysis constants.[33] In fact, while most metal triflates are prepared under strict anhydrous conditions, lanthanide triflates are reported to be prepared in aqueous solution.[34] The Friedel–Crafts acylation of substituted benzenes proceeds smoothly in the presence of catalytic amount (10% mol) of lanthanide triflates.[35] When acetic anhydride (AAN), acetonitrile, or nitromethane is used as a solvent, the homogeneous acylation reaction readily proceeds, and nitromethane gives the highest yield of the desired product. On the other hand, in carbon disulfide, methylene chloride, or nitrobenzene, the reaction mixture is heterogeneous, and the yields of the acylation product are very poor.

Para-acetylation of anisole with AAN is performed in the presence of ytterbium triflate (20% mol). The reaction proceeds smoothly even when the catalyst is employed in a small amount (5% mol), and the desired acylation product is obtained in 79% yield. Several substituted benzenes are subjected to ytterbium-triflate-catalyzed acetylation. Though acetylation of benzene does not occur, introduction of a methylthio or dimethylamino group on the aromatic ring gives the acylation products in high yields. The presence of a methyl group is less effective, and *meta*-xylene is acylated to 2,4-dimethylacetophenone in only 25% yield. Ytterbium triflate can be recovered from the aqueous layer by simple extraction and reused for two additional runs in the model acetylation of anisole, affording *para*-acetylanisole with almost the same yield as in the first use.

Very interesting results can be achieved when scandium triflate is utilized instead of ytterbium triflate.[36] Indeed, acylation of methoxy- and

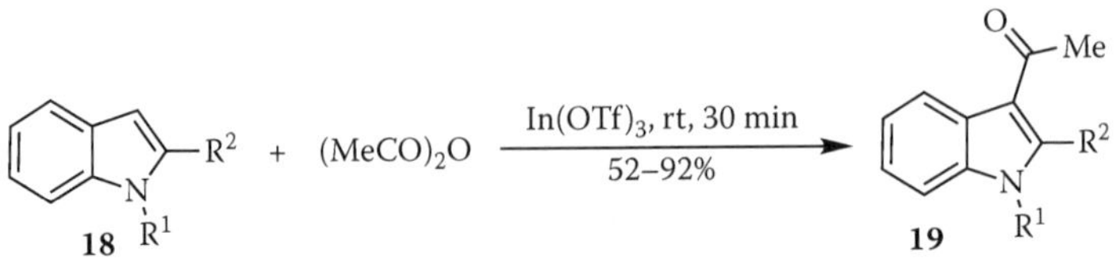

R^1 = H, SO$_2$Ph, tosyl
R^2 = H, Me

Scheme 3.4

thiomethoxy arenes with aliphatic and aromatic anhydrides or acyl chlo-rides gives the corresponding ketones in 79%–99% yield by using scan-dium triflate (20% mol) in nitromethane as solvent. The catalyst can again be recovered and reused giving similarly good results.

Scandium triflate has been patented as an efficient homogeneous cata-lyst in the electrophilic acylation of aromatic compounds with carboxylic acids. 1-Naphthol reacts with AAC at 100°C in the presence of little scandium triflate for 6 h, giving 2-acetyl-1-naphthol in 54% yield.[37]

2-Substituted indoles **18** undergo easy 3-acetylation with AAN in the presence of indium triflate (5% mol) at room temperature, affording prod-ucts **19** (Scheme 3.4).[38] Reactions are performed without solvent by using an excess of AAN.

The efficiency of lanthanide triflates in Friedel–Crafts acylation can be improved by performing the reaction in a lithium perchlorate–nitromethane solution.[39] Among the Lewis acids screened, hafnium triflate is shown to be the most effective in the acylation of mesitylene with AAN. The turnover number (TON) of different triflate and halide catalysts is 5–20 times improved in 4 M lithium perchlorate-nitromethane. The reaction catalyzed by hafnium triflate (5% mol) under the same condi-tions is efficiently applied to different aromatic substrates. Even sterically hindered anhydrides such as benzoic, isobutyric, and pivalic anhydrides give good yields (83%–99%). It is assumed that, in the model Friedel–Crafts acylation with AAN, an active species can be generated with lithium per-chlorate in the presence of a catalytic amount of hafnium triflate. Although the precise structure is not yet clear, IR data support the existence of this species. The absorption of the carbonyl groups of AAN in nitromethane is recognized at 1826 and 1757 cm^{-1}, while absorptions are observed in lower wavenumbers in the presence of hafnium triflate whose intensities depend on the amount of lithium perchlorate.[40]

When *meta*-xylene is acylated in the presence of scandium triflate (20% mol) combined with lithium perchlorate in nitromethane, 2,4-dimethyl-acetophenone is obtained in 89% yield. In the absence of scandium triflate, lithium perchlorate is not soluble in the reaction mixture, and no acylation

Table 3.7 Effect of lithium perchlorate addition on product distribution in the reaction between 2-MN and AAN catalyzed by antimony(III) triflate

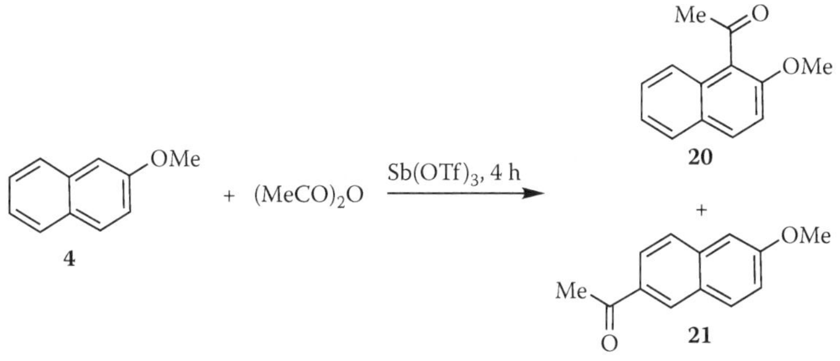

$$\text{4} \quad (\text{OMe}) \quad + \quad (MeCO)_2O \quad \xrightarrow{Sb(OTf)_3,\ 4\ h} \quad \text{20} \quad + \quad \text{21}$$

				Product distribution (%)	
Additive	Solvent	$T\ (°C)$	Yield (%)	**20**	**21**
—	MeNO$_2$	rt	81	95	5
—	MeCN	50	85	98	2
—	CH$_2$Cl$_2$	50	70	94	6
LiClO$_4$	MeNO$_2$	50	93	0	100

product is obtained. Moreover, the catalyst can be recovered in 94%–96% yield and reused for at least three successive runs, giving the product with only a small lowering of yield.[41]

A remarkable effect of the addition of lithium perchlorate is found in the acetylation of 2-MN **4** with AAN in the presence of antimony(III) triflate (5% mol).[42] Indeed, when acetylation is performed in nitromethane at room temperature, or in acetonitrile or methylene chloride at 50°C without lithium perchlorate, product **20** is obtained preferentially (Table 3.7). On the other hand, a dramatic change of regioselectivity is observed when the same reaction is performed in nitromethane/lithium perchlorate, and compound **21** is obtained in 93% yield. The 1-acetylated kinetic adduct **20** is formed in the initial stage of the reaction, and the migration from **20** to the thermodynamic 6-acetylated adduct **21** occurs during the reaction (as described in detail in Chapter 4).

Experimental conditions were developed that require the use of indium(III) triflate (1% mol) in the acetylation of anisole with AAN in combination with silver perchlorate in acetonitrile at 50°C.[43] The reaction gives *para*-methoxyacetophenone in 82% yield, and better results, namely, 96% yield, can be achieved by increasing the lithium perchlorate amount (from 25% to 100%) in nitromethane at 50°C for 1 h. The catalytic system is highly effective for the acetylation of electron-rich aromatic and

Table 3.8 Acetylation of (hetero) aromatics with AAN in
the presence of indium(III) triflate–lithium perchlorate
mixture in nitromethane at 50°C for 1 h

Aromatic substrate	Catalyst amount (%)	Product	Yield (%)
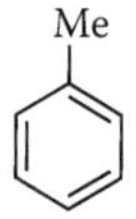	10	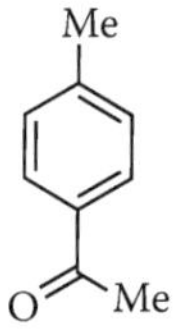	82
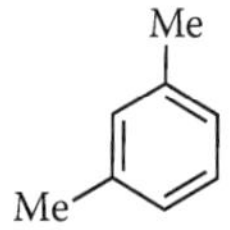	1	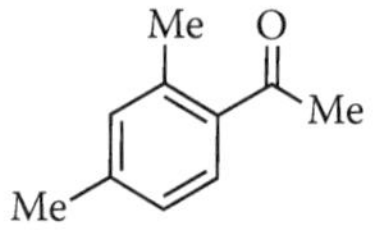	90
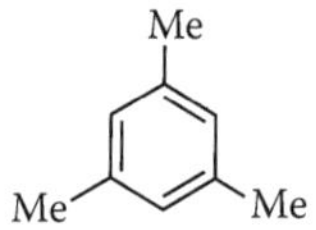	1	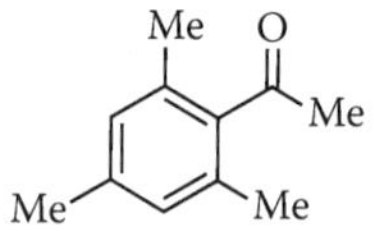	99
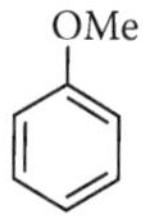	1	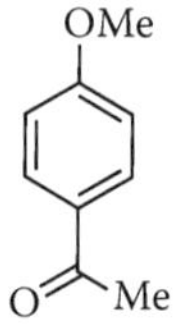	96
	1	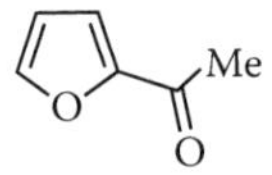	99

heteroaromatic compounds (Table 3.8). Unfortunately, the reaction does not proceed cleanly with unactivated aromatics.

Intramolecular Friedel–Crafts acylation of enolizable benzyl Meldrum acids **22** with scandium triflate (7%–12% mol) in refluxing nitromethane or acetonitrile is a powerful tool for the preparation of a variety of indan-1-ones **23** (Table 3.9).[44,45] Meldrum acid derivatives, mono- and dis-ubstituted at the benzylic position, are accessed via 1,4-conjugate addition of aryl Grignard reagents to Meldrum's alkylidenes, which in turn are prepared by Knoevenagel condensation of Meldrum's acid with ketones under basic or acid catalysis. The best solvent for the cycloacylation **22→23** is nitromethane at 100°C, and the more efficient catalyst is scandium tri-flate (7%–12% mol) for 15 to 560 min. The mildness of the acylation protocol is compatible with acid-labile functional groups such as 1,3-dioxolane.

Table 3.9 Preparation of variously substituted indan-1-ones
catalyzed by scandium triflate

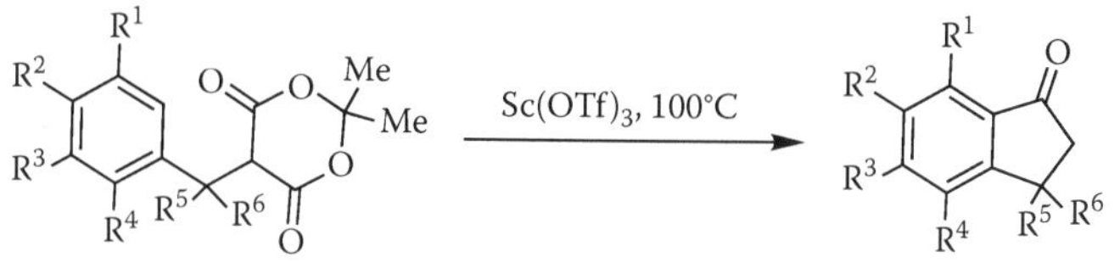

R¹	R²	R³	R⁴	R⁵	R⁶	Solvent	t (min)	Yield (%)
OMe	H	OMe	H	H	H	MeNO$_2$	60	73
OMe	H	OMe	H	H	Me	MeCN	120	77
OMe	H	OMe	H	Me	Me	MeCN	120	82
OMe	H	OMe	H	-(CH$_2$)$_5$-		MeCN	120	83
Me	H	Me	H	H	H	MeNO$_2$	560	52
Me	H	Me	H	-(CH$_2$)$_5$-		MeNO$_2$	90	75
H	OMe	OMe	H	H	H	MeNO$_2$	45	59
H	OMe	OMe	H	H	Me	MeNO$_2$	20	68
H	OMe	OMe	H	Me	Me	MeNO$_2$	20	69
OMe	H	H	OMe	H	H	MeNO$_2$	20	32
OMe	H	H	OTBDPS	H	H	MeNO$_2$	35	63
H	H	OMe	OMe	H	H	MeNO$_2$	80	64
H	H	OMe	OMe	H	Me	MeNO$_2$	40	38
H	H	OMe	OTBDPS	H	H	MeNO$_2$	40	50
H	H	OMe	OTIPS	H	H	MeNO$_2$	40	45
H	H	H	H	H	H	MeNO$_2$	540	13
H	H	H	H	-(CH$_2$)$_5$-		MeNO$_2$	30	56
H	H	H	H	Me	(CH$_2$)$_5$Cl	MeNO$_2$	15	52

The reaction can also be extended to quaternized benzyl Meldrum acids **24**, affording products **25** with a different substitution on the cyclopentanone ring (Table 3.10). The method is also applied to the synthesis of 1-tetralones and 1-benzosuberones.

Ionic liquids can provide an ideal medium for reactions that involve reactive ionic intermediates due to their ability to stabilize charged intermediates such as acyl cations in Friedel–Crafts acylation.[46] As shown in the previous chapter, examples of electrophilic aromatic acylation are reported utilizing ionic liquid-catalysts based on classic Lewis acids, namely, [emim]Cl-aluminum chloride[47] and [emim]Cl-iron trichloride, which can give acylation of both activated and deactivated aromatic compounds.[48]

Metal triflates that do not require special handling are also utilized as catalysts in the Friedel–Crafts acylation performed in ionic liquids with

Table 3.10 Preparation of variously substituted indan-1-ones promoted by scandium triflate

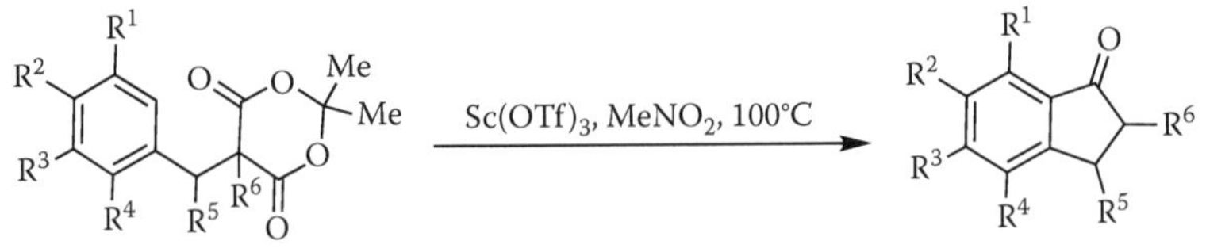

R^1	R^2	R^3	R^4	R^5	R^6	t (min)	Yield (%)
OMe	H	OMe	H	H	Me	45	80
Me	H	Me	H	H	Me	30	87
H	OMe	OMe	H	H	Me	45	77
H	OMe	OMe	H	H	Ph	60	67
H	OMe	OMe	H	H	Bn	45	80
H	OMe	OMe	H	H	CH$_2$CH=CH$_2$	45	76
H	OMe	OMe	H	H	CH$_2$C≡CH$_2$	45	80
H	OMe	OMe	H	H	4-CN-C$_6$H$_4$CH$_2$	320	78
H	OMe	OMe	H	H	4-NO$_2$-C$_6$H$_4$CH$_2$	250	81
H	OMe	OMe	H	H	PhF$_5$-CH$_2$	85	80
OMe	H	OMe	H	Me	Me	155	91
OMe	H	OMe	H	Me	Bn	1935	92
OMe	H	H	OMe	H	Me	20	69
OMe	H	H	OTBDPS	H	Me	20	94
H	H	OMe	OMe	H	Me	20	75
H	H	OMe	OTBDPS	H	Me	30	86
H	H	OMe	OTIPS	H	Me	40	77
H	H	H	H	H	Bn	50	66

some advantages compared to conventional solvents.[49] In the benzoylation of anisole with BC in [bmim][BF$_4$] in the presence of copper(II) triflate (10% mol) at 80°C, 100% conversion BC is achieved with an ortho/para ratio of 0.04. The reaction can be extended to different methoxyarenes and mesitylene with different aliphatic and aromatic acyl chlorides and anhydrides (48%–91% yield). Some cases merit special comment: the benzoylation of 2-MN with BC exclusively produces the kinetic product 1-benzoyl-2-methoxynaphthalene in 72% yield, which cannot be isomerized in [bmim][BF$_4$] into the thermodynamic isomer 6-benzoyl-2-methoxynaphthalene even after prolonged reaction times, reminiscent of the acylation of naphthalene with acetyl chloride (AC) in ionic liquid-catalyst [emim]Cl-aluminum

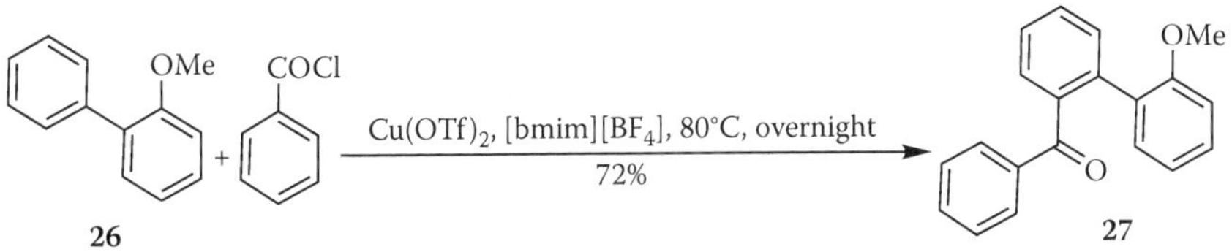

Scheme 3.5

chloride (emim = 1-methyl-3-ethylimdazolium cation) where the major product is the thermodynamically unfavored 1-isomer with 98/2 1-isomer/2-isomer molar ratio.[11] It is important to underline that, if naphthalene acetylation is carried out in nitrobenzene, the 2-isomer is the major product,[50] and the acylating agent is thought to be the free acylium ion.[51] Since the acylium ion is much smaller than the adduct, attack at the sterically more hindered position can occur.

The benzoylation of 2-methoxybiphenyl **26** affords the substitution product 2-(2-methoxyphenyl)benzophenone **27** in copper(II) triflate/[bmim] [BF₄], whereas traditional benzoylation with aluminum chloride, occurring exclusively at the position para to the methoxy group, is reported in the literature (Scheme 3.5).[52]

In the benzoylation of anisole with BC, the copper(II) triflate/[bmim] [BF₄] is recycled three times after washing with diethyl ether, affording the methoxy-substituted BPs in 86%, 79%, and 65% yield with an ortho/para ratio of 0.04 for all runs. The lowering of the catalyst efficiency is assumed to be due to the production of hydrogen chloride, possibly leading to the formation of an increasing amount of copper dichloride, which is inactive under these conditions.

Bismuth(III) triflate and bismuth(III) oxide immobilized in ionic liquids are efficiently utilized in the acylation of aromatics with acyl chlorides.[53] In the reaction of anisole with BC catalyzed by bismuth(III) triflate (10% mol) and carried out in neat anisole, 92% isolated yield of BPs (ortho/para = 0.09) is obtained after 3 h at 110°C. Quite remarkably, bismuth(III) triflate (1% mol), dissolved in [emim][NTf₂], gives complete conversion of BC at 80°C after 4 h. The higher activity of these bismuth(III) derivatives in ionic liquids is further confirmed in the benzoylation of the less reactive toluene with BC. At 150°C in the presence of [emim][NTf₂], 100% BC conversion is observed after 3 h. The bismuth(III) oxide/[emim] [NTf₂]-catalyzed benzoylation can be extended to different activated aromatic compounds, affording ketones with 75%–91% yield. Benzoylation of benzene or chlorobenzene gives the corresponding BPs in 88% and 91% yield, respectively, in the presence of bismuth(III) triflate/[emim][NTf₂]. The catalyst can be recycled, showing the same activity after four runs.

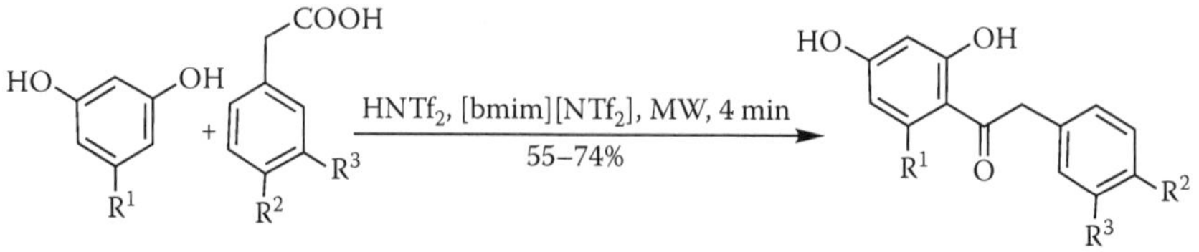

R^1 = H, OH
R^2 = OH, OMe
R^3 = H, OMe

Scheme 3.6

Comparable yields are also obtained in the acylation of activated aromatic compounds with carboxylic acids using a substoichiometric amount of bis[(trifluoromethyl)sulfonyl]amine in [emim][NTf$_2$].[54] Thus, polyhydroxy-deoxybenzoins 30 can be synthesized in 55%–74% yield by acylation of polyhydroxylated aromatic compounds 28 with suitable phenylacetic acids 29 (Scheme 3.6).

Ionic liquids, in combination with supercritical fluids, have been shown to be versatile media for the immobilization and continuous recy-cling of homogeneous catalysts.[55] Thus, 1-butyl-4-methylpyridinium bis-triflimide ([4-MBP][NTf$_2$]) represents a suitable ionic liquid as it shows negligible mass loss when extracted with supercritical carbon dioxide and apparently no negative influence on metal triflate catalytic activity. The anisole acylation with AAN can be performed continuously using a metal triflate catalyst (5% mol) immobilized in the ionic liquid and supercritical carbon dioxide as the continuous extraction phase.[56] Among the salts tested, yttrium triflate possesses the best balance between sufficient acid-ity for catalytic activity and softness to release the product [half-life: 50 h; maximum conversion: 60%; TON: 177; turnover frequency (TOF): 3.0 h^{-1}].

Direct acylation of activated aromatic compounds with carboxylic acids[57] can be achieved by using metal triflates or bis(trifluoromethanesulfonyl) amide salts (bistriflimide salts) such as scandium bistriflimide,[58] ytterbium(III) bistriflimide,[59] europium(III) bistriflimide,[60] and bismuth(III) bistriflimide.[61] Reactions are carried out at high temperature (180°C–250°C) without solvent in a sealed glass tube, affording higher yields in aromatic ketones than classic Lewis acids; however, experimental conditions must be accurately optimized since the high temperature and the very hard Lewis acidity of the catalyst seems to promote undesirable side reactions. The process can be applied to a great number of aromatic substrates and aliphatic carboxylic acids, affording ketones with good yields by using europium(III) bistriflimide (15% mol) (Table 3.11).

Table 3.11 Europium-bistriflimide-catalyzed acylation of aromatics with aliphatic carboxylic acids

R^1	R^2	R^3	R^4	R^5	t (h)	Yield (%)	Isomer ratio
Me	H	H	H	C_6H_{13}	17	78	o/m/p = 22/5/73
Me	H	H	H	Pr^i	24	58	o + m/p = 14/86
Me	H	H	H	c-C_6H_{11}	24	58	o + m/p = 12/88
Pr^i	H	H	H	C_6H_{13}	19	71	o + m/p = 15/85
Me	H	Me	H	C_6H_{13}	12	74	96/4[a]
Me	Me	H	H	C_6H_{13}	12	85	88/12[b]
Et	H	H	Et	C_6H_{13}	12	74	
Me	H	H	Me	C_7H_{15}	12	80	
Me	H	H	Me	C_8H_{17}	12	86	
Me	H	H	Me	C_9H_{19}	13	82	
H	H	H	H	C_6H_{13}	24	4	
OMe	H	H	H	C_6H_{13}	6	87	o/p = <3/>97
OMe	H	H	H	Pr^i	16	77	o/p = 0/100
OMe	H	H	H	c-C_6H_{11}	16	80	o/p = 0/100

[a] 1-(2,4-Dimethylphenyl)-1-heptanone/other regioisomers.
[b] 1-(3,4-Dimethylphenyl)-1-heptanone/1-(2,3-dimethylphenyl)-1-heptanone.

Similar high yields are obtained in the cycloacylation of aryl butyric acids in toluene at 180°C and in the presence of bismuth(III) bistriflimide (1% mol).[62] A great variety of tetralones, and chroman- and thiochroman-4-ones, can be synthesized in high yield (80%–98%) by carrying out the reaction in *para*-xylene or chlorobenzene.

Ytterbium(III) tris(perfluoroalkanesulfonyl)methides **34** are effective catalysts (10% mol) for the Friedel–Crafts acylation of arenes with anhydrides.[63] Compounds **34** can be prepared as described in Scheme 3.7. Trimethylsilylmethyl lithium **31** is reacted with commercially available perfluoroalkanesulfonyl fluorides, giving intermediates **32** that can similarly produce tris-perfluoroalkanesulfonyl derivatives **33**. These compounds are converted into the catalysts **34** by reaction with ytterbium oxide. It is shown that the highly fluorinated catalyst (**34**, $R^1 = C_6F_{13}$, $R^2 = C_8F_{17}$) (10% mol) gives acylation of anisole with AAN in 82% conversion in fluorous solvents such as perfluorodecalin. The catalyst can be recovered in 96% yield by extraction of the reaction mixture with hot perfluorodecalin. In the

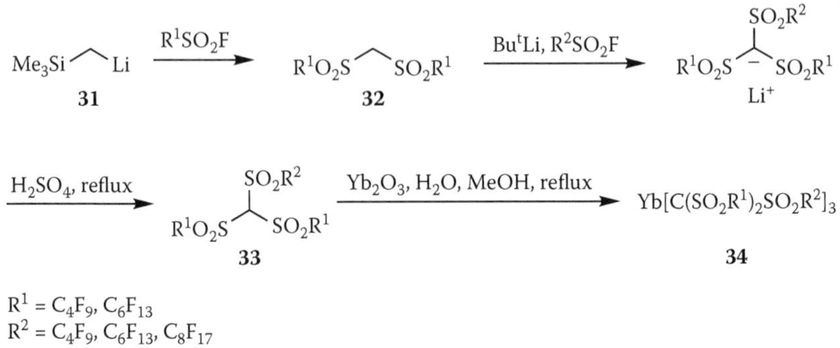

$$Me_3Si \diagup Li \xrightarrow{R^1SO_2F} R^1O_2S \diagup SO_2R^1 \xrightarrow{Bu^tLi, R^2SO_2F} R^1O_2S \diagdown SO_2R^1$$

31   32

$$\xrightarrow{H_2SO_4, reflux} R^1O_2S \diagdown SO_2R^1 \xrightarrow{Yb_2O_3, H_2O, MeOH, reflux} Yb[C(SO_2R^1)_2SO_2R^2]_3$$

33   34

$R^1 = C_4F_9, C_6F_{13}$
$R^2 = C_4F_9, C_6F_{13}, C_8F_{17}$

Scheme 3.7

second run with the recovered catalyst, a 77% conversion to the acylated product is observed.

A great improvement in the catalytic activity can be achieved by performing the acylation of anisole with AAN in trifluoromethylbenzene at 150°C in the presence of catalyst **34** ($R^1 = R^2 = C_4F_9$) (1% mol). Products are obtained in 80% isolated yield with 97% para-isomer. The heating leads to significant product–catalyst dissociation and thereby facilitates catalyst turnover. Different activated aromatics and heteroaromatics can be acylated in high yield under these conditions.[64]

Gallium nonafluorobutanesulfonate (nonaflate) [Ga(ONf)$_3$] (5% mol) shows a very high activity in the acylation of both activated and deactivated aromatic compounds with acyl chlorides without solvent.[65] Some applications are shown in Table 3.12.

Table 3.12 Benzoylation of aromatics with BC catalyzed by gallium nonaflate

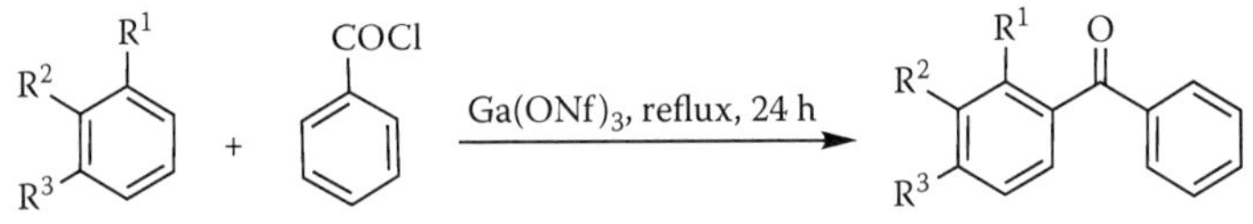

R^1	R^2	R^3	Yield (%)
Me	H	Me	95
H	H	Me	77
H	H	H	90
H	H	F	91
H	H	Cl	89
H	Cl	Cl	65

Metal bistriflimides, already described as efficient catalysts for acylation at high temperature,[61] can be more advantageously utilized in combination with ionic liquids. In the acylation of toluene with BC at 110°C without solvent, metals that are not conventionally thought to be Friedel–Crafts catalysts give the best results. This is particularly the case with $M(NTf_2)_2$ [M = manganese(II), cobalt, nickel, and lead(II)], which give results significantly better than either lanthanide bistriflimides or triflates, namely, 99% yield in 3–6 h. When this reaction with a cobalt or nickel bistriflimide (5% mol) is performed in [emim][NTf_2], the reaction time is reduced to 0.5 and 1 h, respectively, and the catalyst/ionic liquid combination can be recycled with 95% yield on the third cycle. Even deactivated aromatic compounds such as chlorobenzene are acylated with BC in the presence of cobalt bistriflimide (5% mol) in [bmim][NTf_2] at 130°C.[66,67]

Some further studies still deal with the Friedel–Crafts acylation in fluorous fluids. These fluids all have very unusual properties such as high density and high stability, low solvent strength and extremely low solubility in water and organic compounds, and, finally, nonflammability. These properties allow their easy handling and reuse. Friedel–Crafts acylation of electron-rich aromatic substrates can be very efficiently performed in a fluorous biphasic system (FBS), which represents a benign technique for phase separation, and catalyst immobilization and recycling.

Acylation of anisole with AAN is carried out in a mixture of GALDEN® SV 135, a suitable fluorous solvent, and chlorobenzene in the presence of hafnium tetra[bis(perfluorooctanesulfonyl)amide] (1% mol) at 70°C–120°C for 1 to 5 h, giving *para*-methoxyacetophenone in 80% yield and 100% selectivity.[68] It is significant to underline that aluminum chloride gives the mentioned product in only 2% yield under the present mild reaction conditions. The catalyst can be very easily recycled by directly reusing the lower fluorous catalytic phase in the successive reaction with another mixture of reactants, affording the product more than three times without decrease in catalytic activity. The catalytic process can be applied to dimethoxybenzenes and mesitylene with both aliphatic and aromatic acyl chlorides and anhydrides (80%–97% yield). Benzoylation of toluene gives *para*-methylbenzophenone in 35% yield.

The benzoylation of *para-tert*-butylanisole can be performed in the presence of zinc chloride (10% mol) in refluxing perfluoro 2-butyltetrahydrofuran (99°C–107°C), giving 2-benzoyl-4-*tert*-butylanisole in 62% yield.[69]

Ytterbium nonafluorobutylsulfonylimide complex $Yb[N(SO_2C_4F_9)_2]_3$ (10% mol) is employed in the Friedel–Crafts acylation of anisole with AAN. The reaction is performed in benzotrifluoride at 40°C for 2 h, affording quantitative yield of the C-acylation product.[70]

Table 3.13 Benzoylation of aromatics with BC in the presence of indium metal

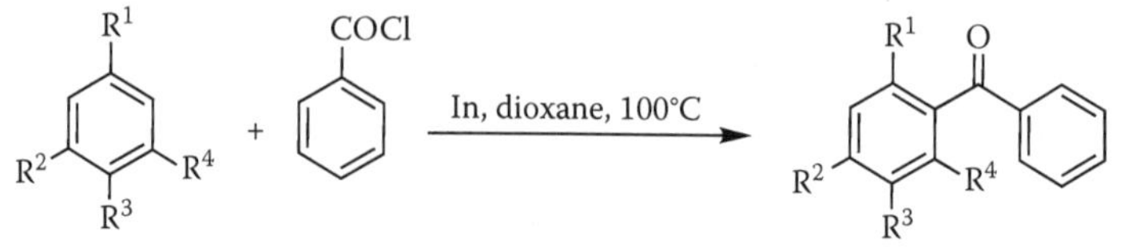

R^1	R^2	R^3	R^4	t (h)	Yield (%)
H	NMe$_2$	H	H	0.75	92
OMe	OMe	H	OMe	1.5	91
OMe	H	OMe	H	2	87
H	SMe	H	H	2	90
H	Me	H	H	2.5	21[a]
Me	H	Me	H	2.5	40[a]
Me	Me	H	Me	0.5	77[a]

[a] Reaction carried out under solventless conditions at reflux.

3.3 Metals

The activity of some metals such as indium, zinc, and iron as efficient catalysts in Friedel–Crafts acylation is well known, even if not clearly explained. For example, indium metal (20% mol) shows high efficiency in the acylation of various aromatic compounds with BC.[71] As an example, the indium-mediated benzoylation of aromatic compounds with electron-donating groups gives the corresponding diaryl ketones in high yield (Table 3.13). Alkylbenzenes are less reactive, and with mesitylene, the yield is 77% at 150°C. Deactivated benzenes such as chloro- and bromobenzene are inactive.

The reaction occurs in a polar and basic solvent such as dioxane; moreover, radical scavengers such as BHT and TEMPO show no negative effect. This implies that the reaction does not proceed via a radical pathway, and the possibility of the participation of indium trichloride generated in situ as a promoter for the reaction cannot be excluded.[43] More likely, the reaction proceeds by an ionic pathway involving an indium(III) complex having an alkyl or aryloxycarbenium ion character[72] that reacts with aromatic compounds to generate the Friedel–Crafts acylation product and indium metal along with hydrogen chloride.

Zinc powder (100% mol) without any treatment promotes Friedel–Crafts acylation of aromatic compounds with acyl chlorides to afford the corresponding aromatic ketones in high yield.[73] Acetylation and benzoylation of a great number of both activated and deactivated aromatic compounds

Table 3.14 Acylation of aromatics with acetyl and benzoyl chlorides in the presence of zinc powder under microwave irradiation

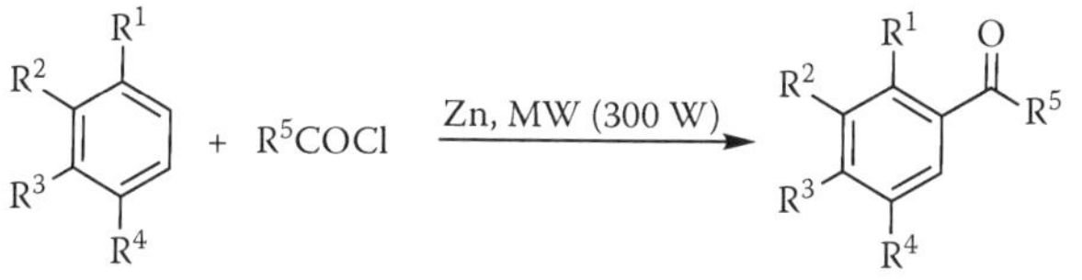

R¹	R²	R³	R⁴	R⁵	T (°C)	t (s)	Yield (%)
H	H	Cl	H	Me	65–68	30	99
H	H	OMe	H	Me	80–82	7	99
H	CH=CH-CH=CH		H	Me	53–55	25	69
H	H	Ph	H	Me	52–54	25	72
OH	H	H	H	Me	55–57	40	73
OH	H	OH	H	Me	64–66	75	77
H	OH	OH	H	Me	44–46	55	71
H	NO$_2$	H	NH$_2$	Me	52–54	20	68
H	NO$_2$	NH$_2$	H	Me	70–72	15	74
H	H	NH$_2$	H	Me	43–45	40	74
H	H	NH$_2$	H	Ph	52–54	20	70
H	NO$_2$	NH$_2$	H	Ph	140–142	30	55
NO$_2$	H	NH$_2$	H	Ph	75–79	15	71
NH$_2$	H	Me	H	Ph	70–72	10	70

can be performed under solventless conditions by irradiation in a microwave oven for 30 s at 300 W. Synthetic results are reported in Table 3.14. Aromatic compounds highly deactivated toward electrophilic substitution such as substituted nitrobenzenes and chlorobenzene undergo acylation in high yield under mild conditions; pyridine affords the corresponding 4-acetylpyridine in 72% yield. It must be underlined that, with phenols and anilines, only C-acylation is observed.

Comparative experiments carried out with anisole and AC confirm that the in-situ-produced zinc chloride has a minor catalytic function, the major role being played by zinc powder, which can be reused up to six times by washing with diethyl ether and hydrogen chloride after each run.

The use of metal in Friedel–Crafts acylation has been patented for the production of diketones, significant in the field of material chemistry. Thus, 1,4-bis(4-methoxybenzoyl)benzene **36** can be synthesized by double acylation of anisole with terephthaloyl chloride **35** in the presence of zinc or iron metal (Scheme 3.8).[74]

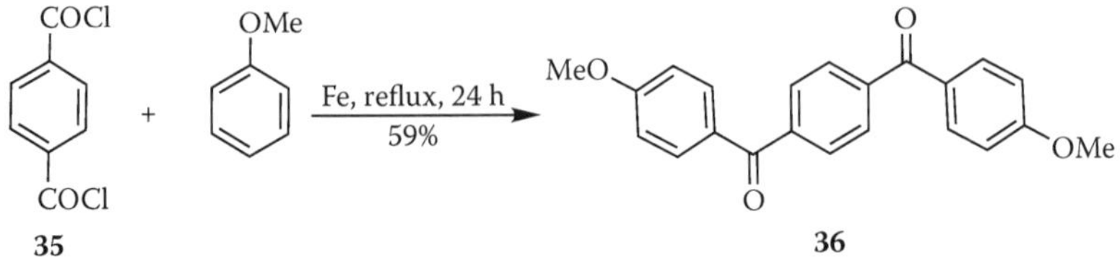

Scheme 3.8

3.4   Miscellaneous

Iodine (2% mol) can be used as catalyst for the acetylation of electron-rich aromatic compounds by aliphatic and aromatic acyl chlorides or anhydrides in 25%–93% yields.[1] In successful acylations, the violet-colored refluxing mixture disappears after 15–30 min. Heterocyclic compounds such as furan and thiophene derivatives undergo easy acylation in the presence of variable amounts of iodine. The process is of particular synthetic interest since these heterocycle compounds are particularly sensitive to acid treatment and give rise to telomerization in classical Friedel–Crafts acylations. Catalysis by iodine is due to the possibility of iodine's acting as an electron acceptor and forming polarized complexes, giving the activate intermediate the possibility of being the type reported in Figure 3.2.[50]

The selective propionylation of benzodioxole **37** at position 5 with propanoic anhydride can be performed in the presence of catalytic amount of aqueous perchloric acid (20% mol) (Scheme 3.9).[75] The reaction is performed in cyclohexane or decalin for 3 h at room temperature. Compound **38**, obtained in 65% yield, represents an intermediate for the industrial production of pyrethroid insecticides. A typical batch reactor for this process is depicted in Figure 3.3.

Halobenzenes are selectively benzoylated at the para-position by a reaction with different BCs in the presence of anhydrous iron(III)sulfate (1% mol).[76] Satisfactory to good yields are obtained by carrying out the reaction at 150°C under solventless conditions (Table 3.15).

Aluminum hydrogensulfate, a very stable and nonhygroscopic solid material that is insoluble in most organic solvents, represents a good catalyst (0.15% mol) for the selective acylation of alkoxybenzenes with aliphatic

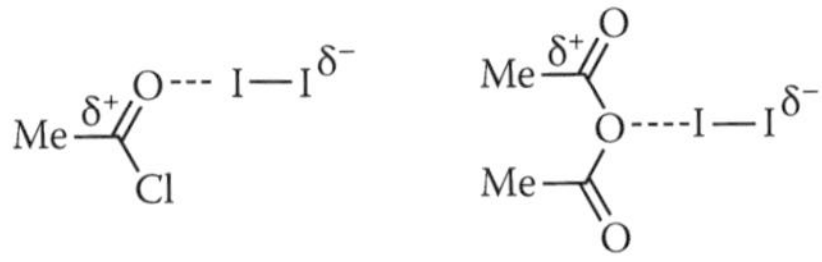

Figure 3.2 Possible AC and AAN activation by iodine.

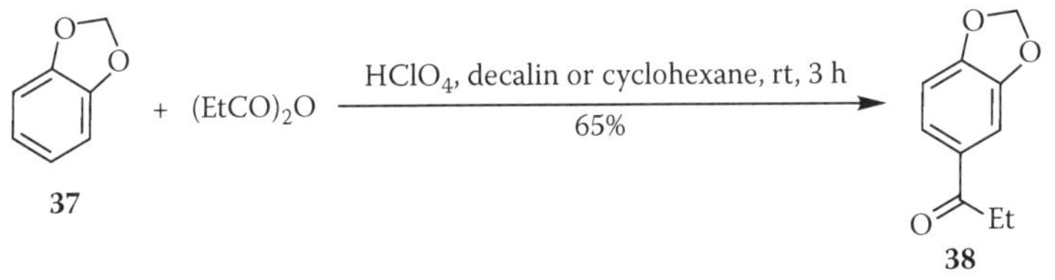

Scheme 3.9

anhydrides (Table 3.16).[77] The reaction is performed without solvent or in nitromethane, giving similar yields.

Bromopentacarbonylrhenium(I) [ReBr(CO)$_5$] promotes the Friedel–Crafts acylation of arenes with acyl chlorides.[78] Toluene undergoes benzoylation with BC in the presence of the rhenium-based catalyst (0.1% mol), affording a mixture of *ortho*-, *meta*-, and *para*-methylbenzophenones in 91% yield (11:4:85 molar ratio). The yield decreases to 72% when using the same catalyst in a lower amount (0.01% mol). The process can be applied to different acyl chlorides, giving the corresponding ketones a satisfactory to high yield (Table 3.17).

Figure 3.3 Batch reactor for the propionylation of benzodioxole with propionic anhydride in the presence of perchloric acid. (From Endura S.p.A. With permission.)

Table 3.15 Ferric-sulfate-catalyzed benzoylation of halobenzenes with BCs

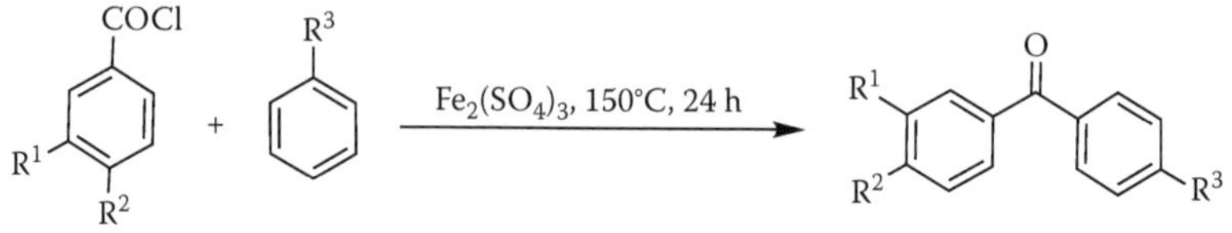

R¹	R²	R³	Conv. (%)	Yield (%)
H	H	F	78	78
H	H	Cl	59	58
H	H	Br	70	78
H	H	I	100	27
H	Cl	Cl	68	69
H	NO₂	Cl	95	79
H	Br	Br	100	63
NO₂	H	Br	96	77
H	Cl	Br	88	78

Table 3.16 Aluminum-hydrogensulfate-promoted
acylation of aromatic ethers with acetic and
propionic anhydrides

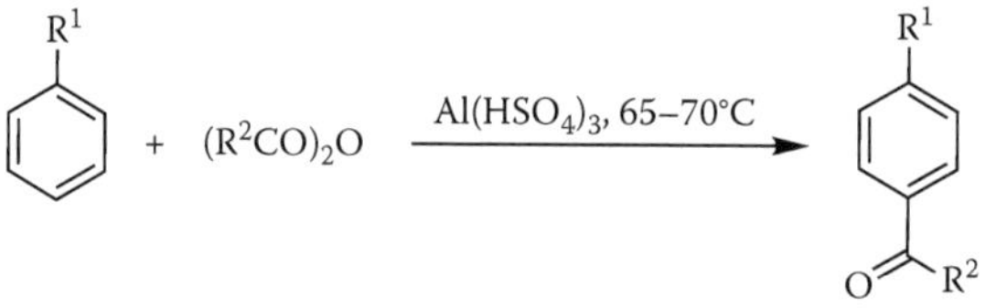

R¹	R²	t (h)	Yield (%)
OMe	Me	1.3	77
OEt	Me	2.1	79
OBu	Me	3.1	75
OBuˢ	Me	5.0	78
OC₆H₁₃	Me	4.5	82
OMe	Et	1.5	78
OEt	Et	2.0	75
OBu	Et	4.0	79
OBuˢ	Et	4.0	82
OC₆H₁₃	Et	3.5	83

Table 3.17 Bromopentacarbonylrhenium(I)-catalyzed acylation of toluene and anisole with acyl chlorides

$$R^1COCl \: + \quad \overset{R^2}{\bigcirc} \quad \xrightarrow{[ReBr(CO)_5], \text{ reflux, 2 h}} \quad \overset{R^2}{\bigcirc}\overset{O}{\underset{R^1}{\diagdown}}$$

R¹	R²	Catalyst (% mol)	Solvent	Yield (%)	o/m/p ratio
Ph	Me	0.1	PhMe	91	11/4/85
PhCH$_2$	Me	0.1	PhMe	84	5/3/92
C$_5$H$_{11}$	Me	0.1	PhMe	73	3/2/95
c-C$_6$H$_{11}$	Me	0.1	PhMe	72	6/2/92
Me	Me	0.1	PhMe	40	6/3/91
Ph	OMe	0.02	ClCH$_2$CH$_2$Cl	92	5/0/95
C$_5$H$_{11}$	OMe	0.02	ClCH$_2$CH$_2$Cl	90	1/0/99
c-C$_6$H$_{11}$	OMe	0.02	ClCH$_2$CH$_2$Cl	95	1/0/99
Ph-CH=CH	OMe	0.02	ClCH$_2$CH$_2$Cl	60	7/3/90

Mechanistic studies have shown that the coordinatively unsaturated [ReBr(CO)$_4$] species, generated by liberation of carbon monoxide from bromopentacarbonylrhenium(I) and/or by the dissociation of the dimeric complex [ReBr(CO)$_4$]$_2$, represents a possible catalytic species.

References

1. Pearson, D. E. and Buehler, C. A. 1972. Friedel–Crafts acylations with little or no catalyst. *Synthesis* 533–542.
2. Effenberger, F. and Steegmuller, D. 1989. Process for the preparation of aryl-(1-phthalimido)alkyl ketones. EP Patent 304,018.
3. Laporte, C., Marquié, J., Laporterie, A., Desmurs, J.-R., and Dubac, J. 1999. Réactions d'acylation sous irradiation micro-onde. II. Acylation d'éthers aromatiques. *C. R. Acad. Sci.* 455–465.
4. Marquié, J., Laporte, C., Laporterie, A., Dubac, J., Desmurs, J.-R., and Roques, N. 2000. Acylation reactions under microwaves. 3. Aroylation of benzene and its slightly activated or deactivated derivatives. *Ind. Eng. Chem. Res.* 39: 1124–1131.
5. Pivsa-Art, S., Okuro, K., Miura, M., Murata, S., and Nomura, M. 1994. Acylation of 2-methoxynaphthalene with acyl chlorides in the presence of a catalytic amount of Lewis acid. *J. Chem. Soc., Perkin Trans. 1* 1703–1707.
6. Carmalt, C. J. and Norman, N. C. 1998. In *Chemistry of arsenic, antimony and bismuth*, ed. N. C. Norman, 1–38. London: Blackie.
7. Répichet, S., Le Roux, C., Roques, N., and Dubac, J. 2003. BiCl$_3$-catalyzed Friedel–Crafts acylation reactions: bismuth(III) oxychloride as a water insensitive and recyclable procatalyst. *Tetrahedron Lett.* 44: 2037–2040.

8. Whitmire, K. H., Labahn, D., Roesky, H. W., Noltemeyer, M., and Sheldrick, G. M. 1991. Sterically crowded aryl bismuth compounds: synthesis and characterization of bis{2,4,6-tris(trifluoromethyl)phenyl} bismuth chloride and tris{2,4,6-tris(trifluoromethyl)phenyl} bismuth. *J. Organomet. Chem.* 402: 55–66.

9. Huang, A.-P., Liu, X.-Y., Li, L.-H., Wu, X.-L., Liu, W.-M., and Liang, Y.-M. 2004. Antimony(V) chloride–benzyltriethylammonium chloride complex as an efficient catalyst for Friedel–Crafts acylation reactions. *Adv. Synth. Catal.* 346: 599–602.

10. Mukaiyama, T., Nagaoka, H., Ohshima, M., and Murakami, M. 1986. The diphenylboryl hexachloroantimonate promoted Friedel–Crafts acylation reaction. *Chem. Lett.* 165–168.

11. Adams, C. J., Earle, M. J., Roberts, G., and Seddon, K. R. 1998. Friedel–Crafts reactions in room temperature ionic liquids. *Chem. Commun.* 2097–2098.

12. da Silveira Neto, B. A., Ebeling, G., Gonçalves, R. S., Gozzo, F. C., Eberlin, M. N., and Dupont, J. 2004. Organoindate room temperature ionic liquid: synthesis, physicochemical properties and application. *Synthesis* 1155–1158.

13. Earle, M. J., Hakala, U., Hardacre, C., Karkkainen, J., McAuley, B. J., Rooney, D. W., Seddon, K. R., Thompson, J. M., and Wähälä, K. 2005. Chloroindate(III) ionic liquids: recyclable media for Friedel–Crafts acylation reactions. *Chem. Commun.* 903–905.

14. Harada, T., Ohno, T., Kobayashi, S., and Mukaiyama, T. 1991. The catalytic Friedel–Crafts acylation reaction and the catalytic Beckmann rearrangement promoted by gallium(III) or an antimony(V) cationic species. *Synthesis* 1216–1220.

15. Mukaiyama, T., Suzuki, K., Han, J. S., and Kobayashi, S. 1992. A novel catalyst system, antimony(V) chloride–lithium perchlorate ($SbCl_5$–$LiClO_4$), in the Friedel–Crafts acylation reaction. *Chem. Lett.* 435–438.

16. Arai, S., Sudo, Y., and Nishida, A. 2005. Niobium pentachloride–silver perchlorate as an efficient catalyst in the Friedel–Crafts acylation and Sakurai–Hosomi reaction of acetals. *Tetrahedron* 61: 4639–4642.

17. Fürstner, A., Voigtländer, D., Schrader, W., Giebel, D., and Reetz, M. T. 2001. A "hard/soft" mismatch enables catalytic Friedel–Crafts acylations. *Org. Lett.* 3: 417–420.

18. Jia, C., Piao, D., Oyamada, J., Lu, W., Kitamura, T., and Fujiwara, Y. 2000. Efficient activation of aromatic C-H bonds for addition to C-C multiple bonds. *Science* 287: 1992–1995.

19. Basavaiah, D., Bakthadoss, M., and Pandiaraju, S. 1998. A new protocol for the syntheses of (E)-3-benzylidenechroman-4-ones: a simple synthesis of the methyl ether of bonduccellin. *Chem. Commun.* 1639–1640.

20. Basavaiah, D. and Reddy, R. M. 2001. One-pot inter- and intramolecular Friedel–Crafts reactions in Baylis–Hillman chemistry: a novel facile synthesis of (E)-2-arylideneindan-1-ones. *Tetrahedron Lett.* 42: 3025–3027.

21. Khodaei, M. M., Alizadeh, A., and Nazari, E. 2007. Tf_2O as a rapid and efficient promoter for the dehydrative Friedel–Crafts acylation of aromatic compounds with carboxylic acids. *Tetrahedron Lett.* 48: 4199–4202.

22. Olah, G. A., Farooq, O., Farnia, S. M. F., and Olah, J. A. 1988. Boron, aluminum, and gallium tris(trifluoromethanesulfonate) (triflate): effective new Friedel–Crafts catalysts. *J. Am. Chem. Soc.* 110: 2560–2565.

23. Desmurs, J. R., Labrouillère, M., Le Roux, C., Gaspard, H., Laporterie, A., and Dubac, J. 1997. Surprising catalytic activity of bismuth (III) triflate in the Friedel–Crafts acylation reaction. *Tetrahedron Lett.* 38: 8871–8874.

24. Izumi, J. and Mukaiyama, T. 1996. The catalytic Friedel–Crafts acylation reaction of aromatic compounds with carboxylic anhydrides using combined catalysts system of titanium(IV) chloride tris(trifluoromethanesulfonate) and trifluoromethanesulfonic acid. *Chem. Lett.* 739–740.

25. Kobayashi, S. and Iwamoto, S. 1998. Catalytic Friedel–Crafts acylation of benzene, chlorobenzene, and fluorobenzene using a novel catalyst system, hafnium triflate and trifluoromethanesulfonic acid. *Tetrahedron Lett.* 39: 4697–4700.

26. Effenberger, F., Eberhard, J. K., and Maier, A. H. 1996. The first unequivocal evidence of the reacting electrophile in aromatic acylation reactions. *J. Am. Chem. Soc.* 118: 12572–12579.

27. Matsushita, Y.-I., Sugamoto, K., and Matsui, T. 2004. The Friedel–Crafts acylation of aromatic compounds with carboxylic acids by the combined use of perfluoroalkanoic anhydride and bismuth or scandium triflate. *Tetrahedron Lett.* 45: 4723–4727.

28. Effenberger, F., Buckel, F., Maier, A. H., and Schmider, J. 2000. Perfluoroalkanesulfonic acid catalyzed acylations of alkylbenzenes: synthesis of alkylanthraquinones. *Synthesis* 1427–1430.

29. Effenberger, F. 1980. Electrophilic reagents—recent developments and their preparative applications. *Angew. Chem., Int. Ed. Engl.* 19: 151–171.

30. Effenberger, F. and Epple, G. 1972. Electrophilic substitution of aromatic compounds. 1. Trifluoromethanesulfonic–carboxylic anhydrides, highly active acylating agents. *Angew. Chem., Int. Ed. Engl.* 11: 299–300.

31. Kobayashi, S. 1994. Rare earth metal trifluoromethanesulfonates as water-tolerant Lewis acid catalysts in organic synthesis. *Synlett* 689–701.

32. Molander, G. A. 1992. Application of lanthanide reagents in organic synthesis. *Chem. Rev.* 92: 29–68.

33. Baes, C. F. Jr. and Mesmer, R. E. 1976. *The hydrolysis of cations.* New York: John Wiley & Sons.

34. Forsberg, J. H., Spaziano, V. T., Balasubramanian, T. M., Liu, G. K., Kinsley, S. A., Duckworth, C. A., Poteruca, J. J., Brown, P. S., and Miller, J. L. 1987. Use of lanthanide(III) ions as catalysts for the reactions of amines with nitriles. *J. Org. Chem.* 52: 1017–1021.

35. Kawada, A., Mitamura, S., and Kobayashi, S. 1993. Lanthanide trifluoromethanesulfonates as reusable catalysts: catalytic Friedel–Crafts acylation. *J. Chem. Soc., Chem. Commun.* 1157–1158.

36. Kawada, A., Mitamura, S., and Kobayashi, S. 1994. Scandium trifluoromethanesulfonate: a novel catalyst for Friedel–Crafts acylation. *Synlett* 545–546.

37. Kobayashi, O. and Hashizume, N. 1998. Production of aromatic ketone. JP Patent 10,087,549.

38. Nagarajan, R. and Perumal, P. T. 2002. $InCl_3$ and $In(OTf)_3$ catalyzed reactions: synthesis of 3-acetyl indoles, bis-indolylmethane and indolylquinoline derivatives. *Tetrahedron* 58: 1229–1232.

39. Hachiya, I., Moriwaki, M., and Kobayashi, S. 1995. Hafnium(IV) trifluoromethanesulfonate, an efficient catalyst for the Friedel–Crafts acylation and alkylation reactions. *Bull. Chem. Soc. Jpn.* 68: 2053–2060.

40. Hachiya, I., Moriwaki, M., and Kobayashi, S. 1995. Catalytic Friedel–Crafts acylation reactions using hafnium triflate as a catalyst in lithium perchlorate–nitromethane. *Tetrahedron Lett.* 36: 409–412.
41. Kawada, A., Mitamura, S., and Kobayashi, S. 1996. Ln(OTf)$_3$-LiClO$_4$ as reusable catalyst system for Friedel–Crafts acylation. *Chem. Commun.* 183–184.
42. Kobayashi, S. and Komoto, I. 2000. Remarkable effect of lithium salts in Friedel–Crafts acylation of 2-methoxynaphthalene catalyzed by metal triflates. *Tetrahedron* 56: 6463–6465.
43. Chapman, C. J., Frost, C. G., Hartley, J. P., and Whittle, A. J. 2001. Efficient aromatic and heteroatom acylations using catalytic indium complexes with lithium perchlorate. *Tetrahedron Lett.* 42: 773–775.
44. Fillion, E., Fishlock, D., Wilsily, A., and Goll, J. M. 2005. Meldrum's Acids as acylating agents in the catalytic intramolecular Friedel–Crafts reaction. *J. Org. Chem.* 70: 1316–1327.
45. Fillion, E. and Fishlock, D. 2003. Convenient access to polysubstituted 1-indanones by Sc(OTf)$_3$-catalyzed intramolecular Friedel–Crafts acylation of benzyl Meldrum's acid derivatives. *Org. Lett.* 5: 4653–4656.
46. Wasserscheid, P. and Keim, W. 2000. Ionic liquids—new solutions for transition metal catalysis. *Angew. Chem., Int. Ed.* 39: 3772–3789.
47. Boon, J. A., Levisky, J. A., Pflug, J. L., and Wilkes, J. S. 1986. Friedel–Crafts reactions in ambient-temperature molten salts. *J. Org. Chem.* 51: 480–483.
48. Davey, N. P., Earle, J. M., Newman, C. P., and Seddon, R. K. 1999. Improvements in or relating to Friedel–Crafts reactions. WO Patent 9,919,288.
49. Ross, J. and Xiao, J. 2002. Friedel–Crafts acylation reactions using metal triflates in ionic liquid. *Green Chem.* 4: 129–133.
50. Olah, G. A. 1973. *Friedel–Crafts chemistry.* New York: Wiley-Interscience.
51. Chevrier, B. and Weiss, R. 1974. Structures of the intermediate complexes in Friedel–Crafts acylations. *Angew. Chem., Int. Ed. Engl.* 13: 1–10.
52. Buu-Hoï, N. P. and Sy, M. 1956. The Pfitzinger reaction with ketones derived from *o*-hydroxydiphenyl. *J. Org. Chem.* 21: 136–138.
53. Gmouh, S., Yang, H., and Vaultier, M. 2003. Activation of bismuth(III) derivatives in ionic liquids: novel and recyclable catalytic systems for Friedel–Crafts acylation of aromatic compounds. *Org. Lett.* 5: 2219–2222.
54. Hakala, U. and Wähälä, K. 2006. Microwave-promoted synthesis of poly-hydroxydeoxybenzoins in ionic liquids. *Tetrahedron Lett.* 47: 8375–8378.
55. Cole-Hamilton, D. J. 2006. Asymmetric catalytic synthesis of organic compounds using metal complexes in supercritical fluids. *Adv. Synth. Catal.* 348: 1341–1351.
56. Zayed, F., Greiner, L., Schulz, P. S., Lapkin, A. and Leitner, W. 2008. Continuous catalytic Friedel–Crafts acylation in the biphasic medium of an ionic liquid and supercritical carbon dioxide. *Chem. Commun.* 79–81.
57. Kawamura, M., Cuiy, D. M., and Shimada, S. 2006. Friedel–Crafts acylation reaction using carboxylic acids as acylating agents. *Tetrahedron* 62: 9201–9209.
58. Ishihara, K., Kubota, M., and Yamamoto, H. 1996. A new scandium complex as an extremely active acylation catalyst. *Synlett* 265–266.
59. Mikami, K., Kotera, O., Motoyama, Y., and Tanaka, M. 1998. Synthesis, structure and high catalytic activity in the Diels–Alder reaction of ytterbium(III) and yttrium(III) bis(trifluoromethanesulfonyl)amide complexes. *Inorg. Chem. Commun.* 1: 10–11.

60. Bhatt, A. I., May, I., Volkovich, V. A., Collison, D., Helliwell, M., Polovov, I. B., and Lewin, R. G. 2005. Structural characterization of a lanthanum bistriflimide complex, $La(N(SO_2CF_3)_2)_3(H_2O)_3$, and an investigation of La, Sm, and Eu electrochemistry in a room-temperature ionic liquid, $[Me_3NnBu]$ $[N(SO_2CF_3)_2]$. *Inorg. Chem.* 44: 4934–4940.

61. Ishihara, K., Hiraiwa, Y., and Yamamoto, H. 2000. Homogeneous debenzylation using extremely active catalysts: tris(triflyl)methane, scandium(III) tris(triflyl)methide, and copper(II) tris(triflyl)methide. *Synlett* 80–82.

62. Cui, D.-M., Kawamura, M., Shimada, S., Hayashi, T., and Tanaka, M. 2003. Synthesis of 1-tetralones by intramolecular Friedel–Crafts reaction of 4-aryl-butyric acids using Lewis acid catalysts. *Tetrahedron Lett.* 44: 4007–4010.

63. Barrett, A. G. M., Braddock, D. C., Catterick, D., Chadwick, D., Henschke, J. P., and McKinnell, R. M. 2000. Fluorous biphase catalytic Friedel–Crafts acylation: ytterbium tris(perfluoroalkanesulfonyl)methide catalysts. *Synlett* 847–849.

64. Barrett, A. G. M., Bouloc, N., Braddock, D. C., Chadwick, D., and Hendersin, D. A. 2002. Highly active ytterbium(III) methide complex for truly catalytic Friedel–Crafts acylation reactions. *Synlett* 1653–1656.

65. Matsuo, J.-I., Odashima, K., and Kobayashi, S. 2000. Gallium nonofluoro-butanesulfonate as an efficient catalyst in Friedel–Crafts acylation. *Synlett* 403–405.

66. Earle, M. J., Hakala, U., McAuley, B. J., Nieuwenhuyzen, M., Ramani, A., and Seddon, K. R. 2004. Metal bis{(trifluoromethyl)sulfonyl}amide complexes: highly efficient Friedel–Crafts acylation catalysts. *Chem. Commun.* 1368–1369.

67. Earle, M. J., McAuley B. J., Ramani, A., Thompson, J. M., and Seddon, K. R. 2002. Metal bis-triflimide compounds, their synthesis and their uses. WO Patent 02/072,260.

68. Hao, X., Yoshida, A., and Nishikido, J. 2005. $Hf[N(SO_2C_8F_{17})_2]_4$-catalyzed Friedel–Crafts acylation in a fluorous biphase system. *Tetrahedron Lett.* 46: 2697–2700.

69. Nakano, H. and Kitazume, T. 1999. Friedel–Crafts reaction in fluorous fluids. *Green Chem.* 1: 179–181.

70. Nishikido, J., Nakajima, H., Saeki, T., Ishii, A., and Mikami, K. 1998. Lanthanide perfluoroalkylsulfonylamide catalysts for fluorous phase organic synthesis. *Synlett* 1347–1348.

71. Jang, D. O., Moon, K. S., Cho, D. H., and Kim, J.-G. 2006. Highly selective catalytic Friedel–Crafts acylation and sulfonylation of activated aromatic compounds using indium metal. *Tetrahedron Lett.* 47: 6063–6066.

72. Inaba, S. and Rieke, R. D. 1985. Metallic nickel-mediated synthesis of ketones by the reaction of benzylic, allylic, vinylic, and pentafluorophenyl halides with acid halides. *J. Org. Chem.* 50: 1373–1381.

73. Satya, P., Puja, N., Gupta, R., and Loupy, A. 2003. Zinc mediated Friedel–Crafts acylation in solvent-free conditions under microwave irradiation. *Synthesis* 2877–2881.

74. Raabe, D. and Hoerhold, H.-H. 1989. Verfahren zur herstellung von 1,4-bis(4-methoxybenzoyl)benzen. DD Patent 272,295.

75. Borzatta, V. and Brancaleoni, D. 2002. Process for the synthesis of 5-(alpha-hydroxyalkyl) benzo[1,3]dioxols. U.S. Patent 6,342,613.

76. Morley, J. O. 1977. Benzoylation of halobenzenes catalysed by iron(III) sulphate. *Synthesis* 54–55.

77. Salehi, P., Khodaei, M. M., Zolfigol, M. A., and Sirouszadeh, S. 2003. Catalytic Friedel–Crafts acylation of alkoxybenzenes mediated by aluminum hydrogensulfate in solution and solvent-free conditions. *Bull. Chem. Soc. Jpn.* 76: 1863–1864.
78. Kusama, H. and Narasaka, K. 1995. Friedel–Crafts acylation of arenes catalyzed by bromopentacarbonylrhenium(I). *Bull. Chem. Soc. Jpn.* 68: 2379–2383.

chapter 4

Catalytic heterogeneous acylations

In this chapter, the use of solid acids as heterogeneous catalysts for the Friedel–Crafts acylation reaction is described. Our review is split up into seven sections, describing the application of zeolites, clays, metal oxides, sulfated zirconia, heteropoly acids, Nafion, and other less-utilized solid catalysts (i.e., graphite). When possible, the relationship between the acid properties of the solids (namely, Brönsted and Lewis types) and the catalytic efficiency is shown, as well as the role of the active site location on the catalyst surface.[1]

However, in some instances, there exist conflicting reports about the strength and nature of the acid sites intervening in the catalysis. These differences may arise not only from the use of different reaction conditions and approaches to preparing or modifying the catalysts but also from a poor characterization of the materials employed. Indeed, the detailed physicochemical characterization of the catalytic materials, as well as the study of their interaction with reagents and products, still represent well-recognized problems in the use of heterogeneous catalysis for organic synthesis, despite the considerable development of spectroscopic techniques observed in the last decade.

4.1 Zeolites

Zeolite-based molecular sieves are crystalline aluminosilicates with a two- or three-dimensional framework consisting of SiO_4 and AlO_4 tetrahedra that are connected by bridging oxygen atoms.[2] Because of the lower valence of aluminum compared to silicon, the number of AlO_4 tetrahedra controls the negative charge on the zeolite framework. This charge is compensated by cations or protons. The protons represent the Brönsted acid sites and participate in the acid-catalyzed transformations of organic molecules. Moreover, dehydroxylation of bridging hydroxy groups leads to the formation of unsaturated aluminum species that act as Lewis acid sites.

Surface acidity, as well as hydrophobicity of zeolites, are usually related to their composition, in particular the silica/alumina ratio (SAR). In the present description, this crucial parameter will be reported between brackets at the end of the zeolite-type code [es: BEA(26)].[3]

Crystalline structures of the zeolite containing tetrahedrically coordinated silicon, aluminum, and phosphorous, as well as transition metals and many group elements with the valence ranging from I to V have also been synthesized with the generic name of zeotypes, including $AlPO_4$-, SAPO-, MeAPO-, and MeAPSO-type molecular sieves.[4,5] They are conventionally defined as ultralarge (>12-), large (12-), medium (10-), or small (8-membered ring) pore materials corresponding to a channel diameter between 20 and 5 Å.

The use of zeolite and zeotype catalysts for organic synthesis[6] shows some advantages compared to conventional catalysts. Some of these advantages are summarized here:

1. A great variety of well-defined crystalline structures differing in channels diameter, geometry, and dimensionality. This favors the so-called shape selectivity, namely, the reactant selectivity that facilitates or prevents the penetration of the reactant molecules into the zeolite channels; product selectivity, which allows only products of appropriate dimensions to diffuse out of the pores; and transition state selectivity, which allows one specific transition state by size or shape effects.
2. Large surface area that can facilitate access of reactants to the active sites.
3. Tunable strength and concentration of the acid sites (both Brönsted and Lewis types), as well as hydrophobicity by isomorphous substitution of trivalent cations (i.e., aluminum, gallium, iron, boron, etc.) into the silicate framework.
4. Making the best catalyst for different chemical reactions by cation exchange using alkali metals, transition metals, and rare-earth metals.

Beta zeolite (BEA) possesses three-dimensional interconnected channels with 12-membered rings and a pore dimension of 0.76 × 0.64 and 0.55 × 0.55 nm.

In the Y zeolite, the spherical internal cavity (the supercage), generated when eight sodalite cages are joined, is about 13 Å in diameter. Entry into the supercage occurs through four identical openings that are 7–8 Å wide, allowing reasonably large molecules to penetrate the internal pores of this three-dimensional material.

Table 4.1 Structural parameters of the most utilized zeolites in Friedel–Crafts acylation reactions

Zeolite	Pore type	Dimension (Å)	Channel dimensionality
BEA	Interconnected channels	7.6 × 6.4/5.5 × 5.5	3D
Y	Interconnected spheres	7.4/11.8	3D
ZSM-5	Interconnected channels	5.3 × 5.6/5.1 × 5.5	3D
MOR	Interconnected channels	6.5 × 7.0/2.6 × 5.7	2D
MCM-41	Parallel channels	15–110	1D

Zeolite ZSM5 is three dimensional and has two types of interconnecting channels: a smaller one (5.1 × 5.5 Å) that is elliptical and linear, and a larger one that is close to circular (5.3 × 5.5 Å).

Mordenite (MOR) zeolite, a bidimensional material, also has two interconnecting channels: the larger one is 6.5 × 7.0 Å, whereas the second one is too small to accommodate organic molecules.

MCM-41 zeotype[7] is defined as mesoporous, and its structure consists of hexagonal, elliptical, or spherical parallel channels that vary from 15 to 110 Å diameter, depending on the templating material chosen for the crystallization process.

In Table 4.1, some structural properties of the zeolites most frequently utilized in Friedel–Crafts acylation reactions are listed. Illustration of the corresponding structures is shown in Figure 4.1.

The metal-exchanged zeolite catalysts are prepared by conventional exchange of zeolites with aqueous solutions of the selected metal nitrate, followed by calcination at the selected temperature before use. In this book, the metal-exchanged zeolites are named MZeolites; for example, the iron-exchanged Y zeolite is named FeY.

Zeolites can also be modified by the dealumination technique. Dealuminated zeolites can be prepared by calcining the starting material impregnated with ammonium fluoride; the treatment increases the SAR framework and reduces the number of Brönsted and Lewis sites.

In the acylation of toluene with isobutyryl chloride or isobutyric anhydride (Scheme 4.1), zeolite activity depends on its structural type and increases from medium- to large-pore zeolites and from the mono- to the three-dimensional channel system.[8] The highest conversion of isobutyryl chloride is obtained with zeolites BEA(25) (71% conversion, 54% *para*-isopropyltolyl ketone yield after 4 h) or ultrastable Y(30) zeolite [USY(30)] (62% conversion after 4 h). When increasing the concentration of the polar acylating agent, which is strongly bound to the zeolite active sites, the reaction rate decreases. It was also observed that the initial rate of acylation with isobutyric anhydride is almost two times higher compared to isobutyryl chloride. Unfortunately, further prolongation of reaction time

Figure 4.1 Structures of the most utilized zeolites in Friedel–Crafts acylation reactions. (From http://izasc.ethz.ch/; http://cpm.tnw.utwente.nl/; http://chemeducator.org/; http://www.zeolyst.com/; http://www.uni-giessen.de/cms/. With permission.)

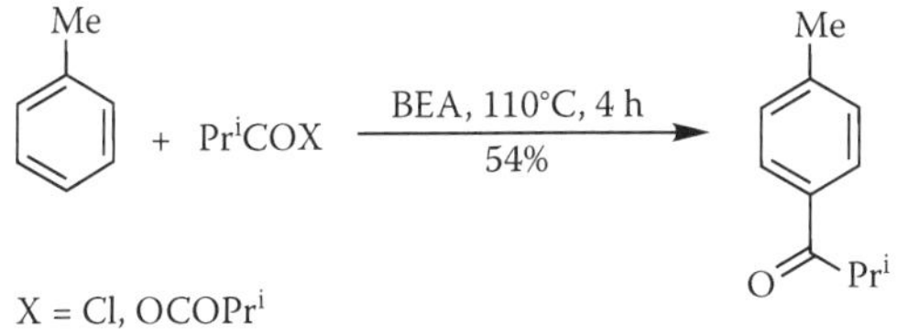

Scheme 4.1

leads to similar final results. However, only one acylating part of isobutyric anhydride can be used; therefore, the utilization of this reagent is not effective.

By decreasing the concentration of acid sites [BEA(75) and BEA(140) series], only a slight decrease of isobutyryl chloride conversion is observed (56% and 53%, respectively). However, the higher turnover frequency (TOF) observed with BEA(140) (3.11 min^{-1}) underscores the importance of the hydrophobicity of zeolites in the reaction between polar and nonpolar compounds. With increasing hydrophobicity of the zeolite, toluene can easily diffuse to the channel system and, also, desorption of polar isopropyltolylketones is much easier. The selectivity toward the *para*-isopropyltolylketone for BEA(25), BEA(75), and BEA(140) is 75%, 78%, and 80%, respectively.

Tridimensional Y and BEA zeolites are also utilized as efficient catalysts in the acylation of tetralin with acyl chlorides.[9] By using octanoyl chloride and butyroyl chloride in the presence of zeolite Y(40), good yields in the acylation products are achieved (83% and 77% yield, respectively, after 8 h), whereas acetyl chloride (AC) leads only to less than 20% yield of the two isomeric ketones **1** and **2** (**1/2** = 7.5) (Scheme 4.2). This trend is correlated to the influence of the chain length of the acylating agent on the efficiency of Friedel–Crafts acylation over zeolite catalysts previously discovered and rationalized.[10] The lower activity of zeolite Y(40) with AC is related to the strong adsorption of organic species, either acetic acid (AAC) or acetyl tetralin.

Butyroyl chloride was utilized in the acylation of thiophene in the presence of zeolite Y[11] with the aim of studying the nature and effect of the catalytic acid sites. The reaction is performed in chlorobenzene at

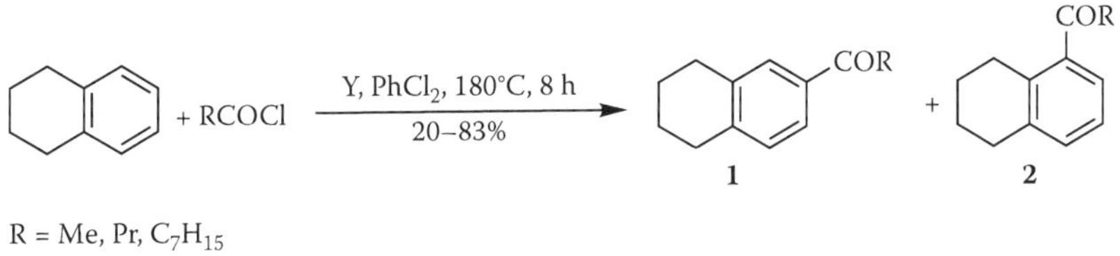

Scheme 4.2

100°C by using USY and dealuminated USY with different SAR values and Brönsted and Lewis site contents. The highest activity is observed in the presence of USY with the maximum Lewis site content. These results indicate an increase in the acylation rate with an increase in the number of Lewis sites. This dependence with respect to the number of Lewis sites is clear, whereas the variation with respect to the number of Brönsted sites is incoherent; this does not mean that the Brönsted sites have no role in the present reaction. The yield in acylated products is generally two orders of magnitude higher than the number of Lewis sites, showing that the catalytic reaction is not stoichiometric in the consumption of the sites.

The role of external or internal surface catalytic sites was also examined with the aim of understanding whether the reaction occurs in the whole volume of the catalyst or only on the external surface. To this end, two USY catalysts with different particle sizes are compared: the first is composed of granules between 28 and 35 mesh, and the other of granules larger than 60 mesh. Both these catalysts exhibit the same activity, suggesting that the process does not take place only on the external surface.

Y(9) zeolite alone, and then Y(9) zeolite treated with nitric acid [Y(56)],[12] were tested in the acylation of *meta*-xylene with benzoic anhydride (BAN), showing similar activity (r_0 = 47 and 54 mol $\cdot$ g^{-1} $\cdot$ min^{-1} $\cdot$ 10^6, respectively);[13] on the contrary, Y(9) shows an appreciably higher catalytic activity (r_0 = 146 mol $\cdot$ g^{-1} $\cdot$ min^{-1} $\cdot$ 10^6) than Y(56) (r_0 = 76 mol $\cdot$ g^{-1} $\cdot$ min^{-1} $\cdot$ 10^6) in the benzoylation of *meta*-xylene with benzoyl chloride (BC). The results indicate that, in Y(9) and Y(56), the catalytically active sites are Brönsted acids. The observed superior activity of Y(9) using BC without solvent is supposed to be due to an interaction of BC or hydrogen chloride with nonframework aluminum species present in the zeolite, resulting in the formation of strong acid sites as evidenced by ^{27}Al NMR spectra. Even if the real nature of nonframework alumina formed in zeolites during activation is still a matter of debate, it appears, however, that this material consists of extremely small particles that have a strong interaction with the zeolite surface.[14]

Polar solvents are known to enhance the acylation reaction over conventional catalysts.[15] Accordingly, sulfolane, a higher polar solvent, enhances appreciably the activity of Y(12) using BC as acylating agent (no solvent: 36%, decane: 20%, and sulfolane: 94% conversion, respectively).

Because of its high localized dipole moment, sulfolane is a very good solvent for inorganic salts, and its effect on the acylation activity of Y(9) is ascribed to the formation of a homogeneous catalyst system consisting of dissolved aluminum compound that interacts with BC. Blank experiments and ^{27}Al NMR analysis of the solution confirm the activity after filtration of the catalyst. These results are not so negative and confirm that a catalytic amount of aluminum species are transferred into solution in sulfolane and catalyzes the quantitative conversion of BC.

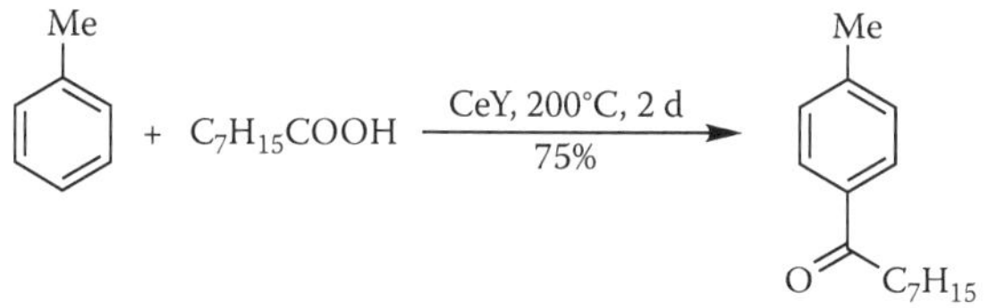

Me

⬡ + C$_7$H$_{15}$COOH → Me–C$_6$H$_4$–C(=O)–C$_7$H$_{15}$

CeY, 200°C, 2 d
75%

O C$_7$H$_{15}$

3

Scheme 4.3

Dealuminated Y(30) zeolite, calcined at 550°C, is found to be a good catalyst for the acylation of veratrole with acetic anhydride (AAN), affording the corresponding 3,4-dimethoxyacetophenone in 95% yield when the reaction is carried out at 90°C for 6 h.[16]

In a series of valuable studies, the activity of various cation-exchanged Y zeolites in the acylation of toluene and xylenes with aliphatic carboxylic acids was investigated.[10,17,18] In a model reaction between toluene and octanoic acid, the activity of rare earth-, transition metal-, and alkaline earth-cation-exchanged Y zeolites was considered.[10] CeY zeolite exhibits the highest activity (**3** yield = 75%) in the para-acylation (Scheme 4.3) in agreement with the results published in an early study;[19] on the contrary, unmodified Y zeolite shows a lower activity (**3** yield < 40%), and transition metal and alkaline-earth-exchanged Y are nearly inactive.

Figure 4.2 shows the dependence of the reaction rate from the cerium content in the CeY zeolite. The catalyst activity increases with the degree of exchange of Na$^+$ ions by Ce^{3+}. There is a threshold of exchange (~20%) below which no activity is observed, but above that threshold a rapid increase in activity occurs. A possible explanation of this phenomenon can be ascribed to the particular structure of the Y zeolite that contains five distinct cation sites that are located, respectively, at the center of the hexagonal prism (type I), in the sodalite unit (types I' and II'), and in the supercage (types II and III)[17] (Figure 4.3).

Concerning the nature of active-site acidic hydroxyl groups, when the zeolite is in the hydrate form, all cations are in the supercage because of the solvation by water. However, upon dehydration, the cation position shifts and, for a low degree of exchange, the trivalent cations tend to migrate toward inaccessible sites I and I', showing very reduced activity. For exchange entities greater than 40%, part of the cations remains in the supercage, and considerable catalytic activity can be observed. The nature of the acid sites and, consequently, the catalyst activity depend on thermal pretreatment. For Y zeolite, the activity remains practically constant for calcination temperatures between 300°C and 600°C; for CeY, the maximum acylation activity is obtained at a calcination temperature of 300°C. This relatively low activation temperature may be a consequence

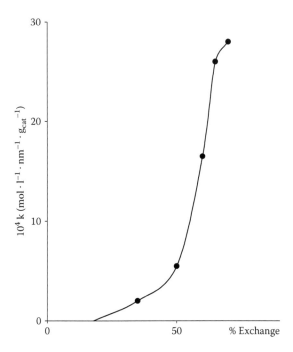

Figure 4.2 Dependence of the degree of sodium ion exchanged by cerium ion on the activity of Y zeolite in the acylation of toluene by octanoic acid at 200°C.

of the easier removal of adsorbed water, and the decrease in activity for temperatures above 350°C is probably due to dehydroxylation of the lattice and conversion of Brönsted acid sites into Lewis ones. These results confirm that Brönsted rather than Lewis acids are the active sites in the present reaction.

A quite interesting behavior is observed by comparing the activity of CeY in the acylation of toluene with linear carboxylic acids of increasing chain length (from acetic [C2] to behenic acid [C22]) (Figure 4.4).[10] Acylation with fatty acids takes place with yields higher than 50%. Acylation of toluene shows an extraordinary high para shape-selectivity effect and, in all cases, the selectivity of the para-isomer is at least 94%, whereas in classical Friedel–Crafts acylation with aluminum chloride, para-selectivity is ~80% due to the competitive attack at the ortho-position. For short-alkyl-chain carboxylic acids, the yield increases linearly with the number of carbon atoms, although all molecules can diffuse through the porous network. This phenomenon is not encountered in homogeneous Friedel–Crafts acylation, and it is probably due to the same reason as previously mentioned, that is, the intracrystalline reaction combined with the hydrophobic interaction of reactants with the catalyst surface.

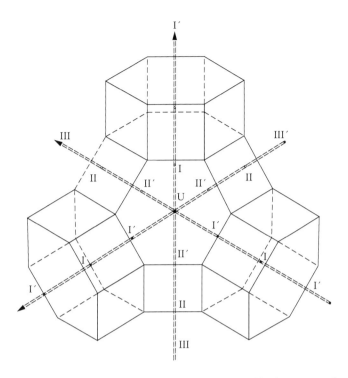

Figure 4.3 Cation sites in Y zeolite. (From Gauthier, C., Chiche, B., Finiels, A., and Geneste, P., *J. Mol. Catal.*, 50, 219, 1989. With permission.)

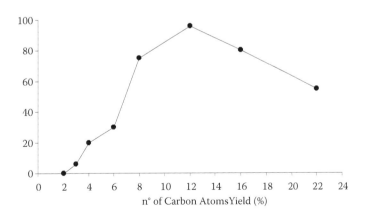

Figure 4.4 Dependence of the yield in toluene acylation with linear carboxylic acids on the chain length in the presence of CeY.

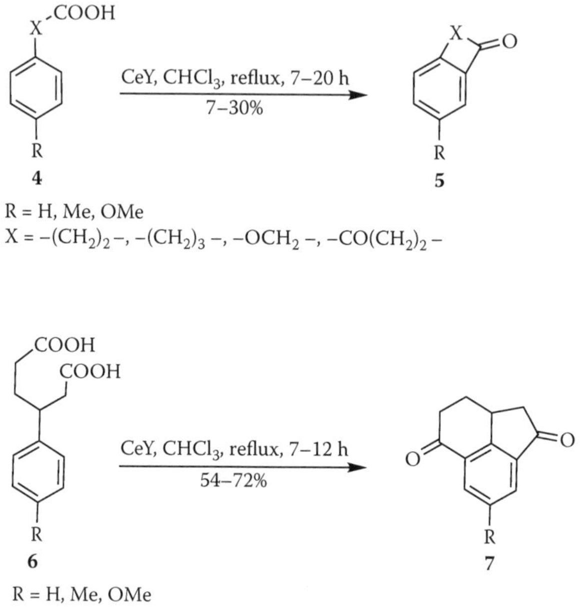

R = H, Me, OMe
X = –(CH$_2$)$_2$–, –(CH$_2$)$_3$ –, –OCH$_2$ –, –CO(CH$_2$)$_2$ –

R = H, Me, OMe

Scheme 4.4

Some synthetic results merit underlining, such as the high-yield and para-selectivity values observed in the acylation of toluene with octanoic, dodecanoic, and palmitic acids that reach values of 75% yield with 94% para-selectivity, 96% yield with 94% para-selectivity, and 80% yield with 99% para-selectivity, respectively.[10,20]

From a synthetic point of view, CeY zeolites are also utilized in the preparation of cycloalkanones via intramolecular electrophilic acylation of variously substituted arylalkanoic acids 4 and 6 (Scheme 4.4).[21] CeY catalyzes the reaction affording bicyclic and tricyclic ketones 5 and 7 in satisfactory to good yield (7%–72%). As well documented in previous studies, the efficiency of the process parallels the lipophilic character of the reagent.

LaY zeolites show good activity in the acylation of anisole as solvent reagent with AC and AAN.[22] The activity of LaY zeolite is dependent on the lanthanum(III) content; increased yield of para-product is found for high levels of lanthanum(III) exchange [60% yield of *para*-acetylanisole with 93% lanthanum(III) exchange].

Cycloacylation of arylpyridine methyl esters 8 to azafluorenones 9 can be efficiently achieved even in the presence of unmodified Y zeolite.[23] The reaction can be performed in cyclohexane at 150°C in a Teflon®-lined autoclave. Synthetic results reported in Table 4.2 confirm the general applicability and efficiency of the method.

Table 4.2 Azafluorenone preparation promoted by Y zeolite

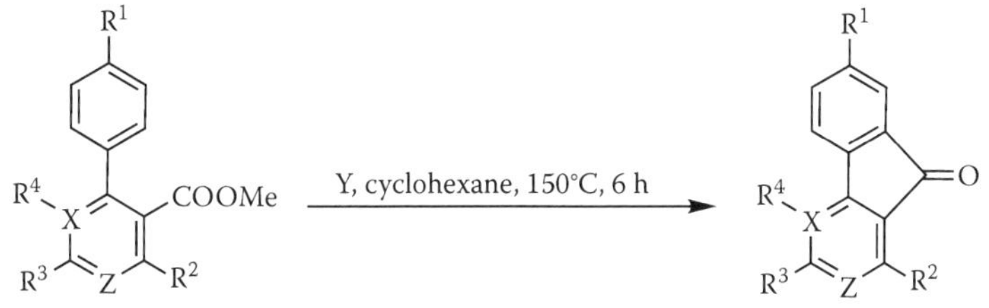

R^1	R^2	R^3	R^4	X	Z	Yield (%)
H	H	H	H	C	N	79
H	H	Me	H	C	N	84
H	H	H	Me	C	N	77
Me	H	H	H	C	N	80
F	H	H	H	C	N	72
Cl	H	H	H	C	N	74
H	H	H	—	N	C	86
F	H	H	—	N	C	76
Cl	H	H	—	N	C	67
H	Me	H	—	N	C	84

The benzoylation of substituted arenes with BC in the presence of FeY was studied in comparison with the homogeneously iron-trichloride-promoted reactions.[24] Results reported in Table 4.3 show that the heterogeneous FeY catalyst gives diaryl ketones in similar yields with respect to the homogeneous counterpart iron trichloride, even if a longer reaction time is required. Comparative experiments confirm that the catalytic activity originates from FeY acting as a truly heterogeneous catalyst and that leaching phenomena are unimportant.

NaY commercial zeolite can be utilized to prepare catalysts with different SAR framework. Y(18) is the best catalyst in the benzoylation of anisole with benzoic acids (BACs).[25] Synthetic results are reported in Table 4.4. The activity of Y zeolite increases with its Lewis acidity, showing that, in this case, the Lewis acid sites are more active than the Brönsted ones. The mechanism of the reaction is found to be similar to that observed under homogeneous catalysis.

The surface acidity role in the acylation of anisole with acyl chlorides was analyzed.[26] From the relative adsorption constants of phenylacetyl chloride and anisole on Y(19) and the absence of any anisole demethylation and rearrangement products, it is possible to conclude that the interaction of the acid sites of the catalyst with anisole oxygen must be not very

Table 4.3 Comparison of the activity of FeCl$_3$ and FeY in the benzoylation of alkylbenzenes with BC

Catalyst	R^1	R^2	t (h)	Yield (%)	Selectivity (%)
FeCl$_3$	H	Me	6	75	82
FeY	H	Me	24	77	80
FeCl$_3$	H	Et	2.5	70	86
FeY	H	Et	24	78	85
FeCl$_3$	H	Bu^t	6	67	91
FeY	H	Bu^t	24	58	91
FeCl$_3$	Me	Me	2.5	91	92
FeY	Me	Me	24	86	92

Table 4.4 Anisole benzoylation with BACs catalyzed by Y zeolite

R^1	R^2	R^3	Conversion (%)	Yield (%)	Selectivity (%)
H	H	OMe	87	80	92
H	H	OH	84	76	90
OH	H	H	89	80	90
H	OH	H	26	19	73
H	H	Me	84	75	89
H	H	H	76	67	88
H	H	Br	60	53	88
H	H	Cl	61	54	89
Cl	H	H	71	65	92
H	Cl	H	54	49	91
I	H	H	82	74	90

strong. On the other hand, the influence of the sodium-exchange level on the initial rate of 1-(4-methoxyphenyl)-2-phenyl ethanone formation shows that the activity of the catalyst is independent on its acid strength distribution. This is confirmed by studying the influence of the SAR framework on the turnover numbers (TONs). The relatively small TON

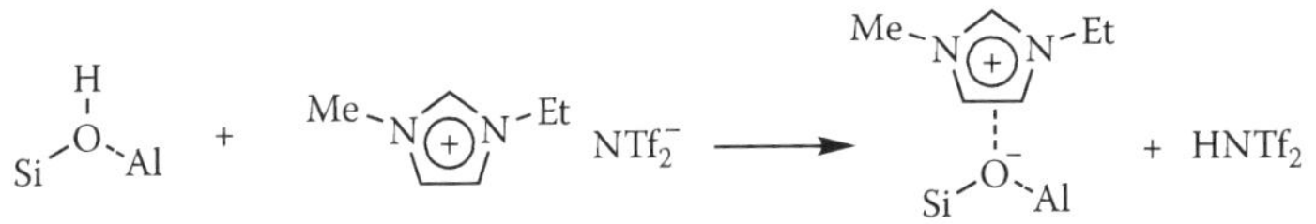

Scheme 4.5

differences can be related to the hydrophobicity of the samples. Similar behavior is observed in the acylation of toluene with acetyl and propanoyl chlorides over LaY zeolites, para-isomers being obtained in 44% and 65% yield, respectively.[27]

The benzoylation of anisole with BAN to give *para*-methoxybenzophenone is studied as a model reaction to evaluate the efficiency of zeolite catalysts in ionic liquids.[28] Higher anhydride conversion is observed by using USY(58) zeolite in ionic liquids (58% in [C_2dmim]NTf$_2$, 55% in [C_2mim]NTf$_2$, and 53% in [C_8py] NTf$_2$) in comparison with classical solvents (30% in 1,2-dichloroethane, 26% in toluene) or under solventless conditions (25%). The reaction is thought to proceed via a homogenous mechanism catalyzed by HNTf$_2$, in situ generated by the exchange of the cation from the ionic liquid with the acid proton of the zeolite (Scheme 4.5). The entire process merits further deepening, in particular to explain the mechanism of catalyst deactivation, because in the reaction carried out in ionic liquids adsorption of the product on the catalyst is not the main source of deactivation.

Acetylation of activated aryl ethers with AAN over different zeolites can be efficiently performed under relatively mild conditions.[29] 1,2-Dimethoxybenzene over BEA(25) and Y(120) gives the corresponding para-regioselective acetylated product in 74% and 91% yield, respectively (Table 4.5). Acetylation of 1,2 diethoxybenzene over the same catalysts affords the para-acetylated product in 78% and 91% yield, respectively.

Under similar conditions, 2,3-dihydrobenzofuran **10** gives 5-acetyl-2,3-dihydrobenzofuran **11** in 95% yield (Scheme 4.6). The solid catalyst can be recovered, reactivated by calcination at 400°C, and reused giving unchanged results.

The activity of some Y zeolites showing different Lewis and Brönsted acid sites density is still studied in the acylation of dimethoxyarenes (with particular attention to veratrole, Scheme 4.7) with different acyl chlorides as a function of zeolite acidity and the lipophilic nature of acyl chlorides.[30] Because of the particularly soft reaction conditions, namely, 65°C and 1 h reaction, 3,4-dimethoxyphenyl ketones **12** are the sole isomers recognized in the reaction mixture. Results of the catalytic tests (yield of compounds **12**) confirm that the best catalyst is Y(14), characterized by an optimum ratio between Lewis and Brönsted acidity (medium-strength acid sites density for Lewis and Brönsted acid = 17.0 and 11.0 mmol × g^{-1} py,

Table 4.5 Acylation of substituted phenyl esters with
AAN in the presence of zeolite catalysts

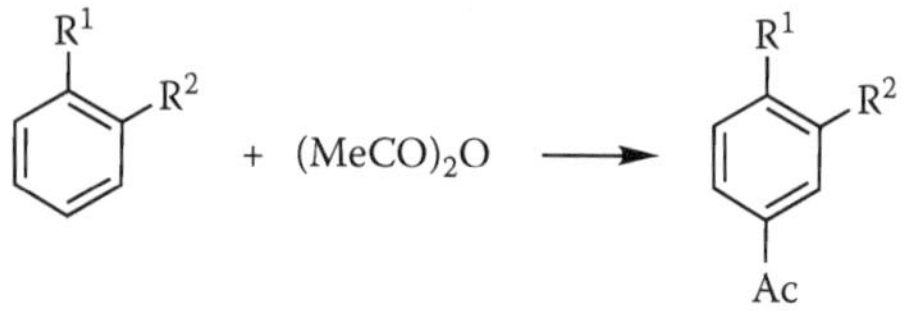

R¹	R²	Catalyst	T (°C)	t (min)	Yield (%)
OMe	OMe	BEA	40	45	74
OMe	OMe	Y	40	45	91
OEt	OEt	BEA	120	60	78
OEt	OEt	Y	120	60	91
OMe	Cl	BEA	120	360	94
OMe	Cl	Y	120	2880	86
OMe	Me	BEA	120	60	87
OMe	Me	Y	120	60	79

respectively), whereas stronger or weaker acid materials give lower yields.
This evidence suggests that the reaction requires both a delicate balance
between the Lewis and Brönsted acid sites and a low density of hydrophilic hydroxy groups that can strongly and preferentially interact with
methoxy and carbonyl groups, resulting in pore blocking by-product
molecules. The chain-length effect of the acyl chloride on ketone yields
shows the typical trend already reported (see Figure 4.4).[10]

The remarkable effect of pore dimensions and structural properties
on zeolite efficiency is further confirmed in the acylation of veratrole

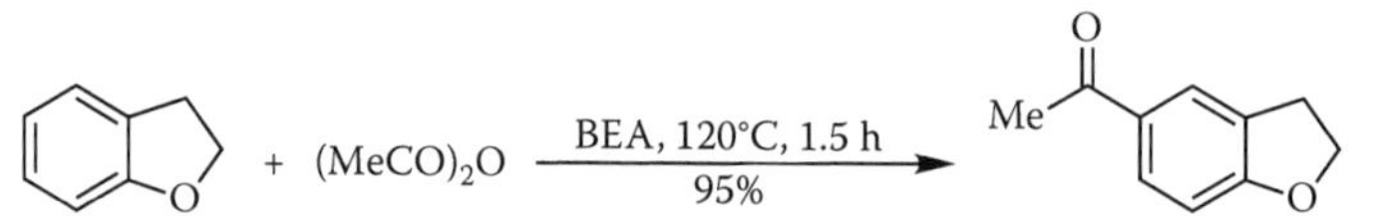

Scheme 4.6

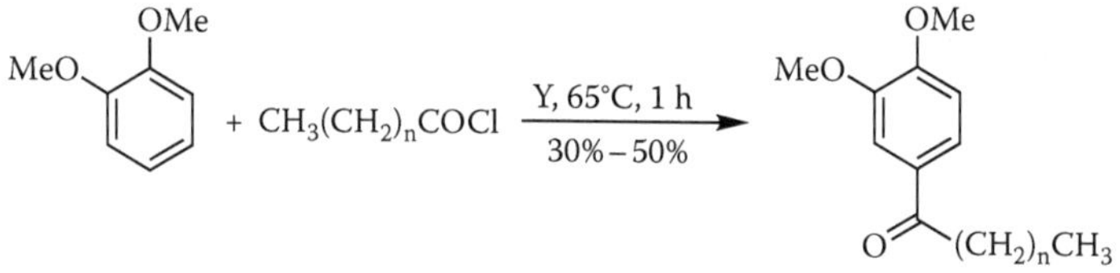

Scheme 4.7

with BAN.[31] The following initial rate value order r_0 (mol $\times$ min^{-1} $\times$ g^{-1}) is observed with some commonly utilized zeolites: Y = 0.125 > BEA = 0.050 > MOR = 0.015. These results can be directly related to the microporous structure of the different catalysts, for which the pores of the tridimensional Y framework allow a readier diffusion of both substrate and product than those of the interconnected channels architecture of BEA and those of the bidimensional MOR framework.

The kinetic study of the reaction of veratrole with AAN over Y(30) zeolite[32] allows for establishing a modified Eley–Rideal type mechanism, where the adsorbed AAN reacts with veratrole in the liquid phase, but in which veratrole is also adsorbed on the active sites of the catalyst, acting in a certain way as a poison of the catalyst. The evidence of a competitive adsorption of both reactants and products has been demonstrated. Thus, a significant inhibiting effect of the reaction product 3,4-dimethoxyacetophenone has been especially shown, leading to a certain loss of activity of the catalyst. Despite such a deactivation, the use of zeolites in the acetylation of dimethoxybenzenes allows the development of a selective procedure for the preparation of the corresponding dimethoxyacetophenones in convenient yields. In particular, the following TOF values are determined in the acetylation of the three isomeric dimethoxybenzenes with AAN over Y(30) zeolite: 1,2-diOMe = 130 min^{-1}; 1,3-diOMe = 130 min^{-1}; and 1,4-diOMe = 49 min^{-1}.

Good synthetic results can be achieved by performing the same reaction in the presence of a dealuminated Y(36) zeolite. Thus, 3,4-dimethoxyacetophenone is obtained in 97% yield by performing the batch reaction at 90°C for 6 h.[33]

Y zeolites also show promising activity in the acylation of heterocyclic compounds. Acylation of benzofurans by AAN was extensively studied, and some good synthetic results have been achieved.[34–37] Benzofuran and 2-methylbenzofuran undergo acylation with AAN under mild conditions, giving 2-acetylbenzofuran **13** and 2-methyl-3-acetylbenzofuran **14** respectively, as the main products over Y(32) zeolite (Scheme 4.8). 2-Methylbenzofuran is about twice as reactive as benzofuran; it is also observed that 2-methylbenzofuran deactivates the catalyst much less than benzofuran. Although this phenomenon is not well understood, it makes possible carrying out the reaction conveniently in a flow reactor. The experimental procedure is found to have a significant influence on the reaction yield: when benzofuran is introduced first, the acylation reaction is practically nonexistent, and very little of the main reaction product **13** is found; when AAN is introduced first, the reaction takes place, and product **13** is obtained in a better yield. The by-products that are entrapped into the catalyst channels are constituted of benzofuran oligomers that can also undergo acetylation reaction. The poisoned catalyst can be regenerated by washing with methylene chloride and calcining at 500°C for

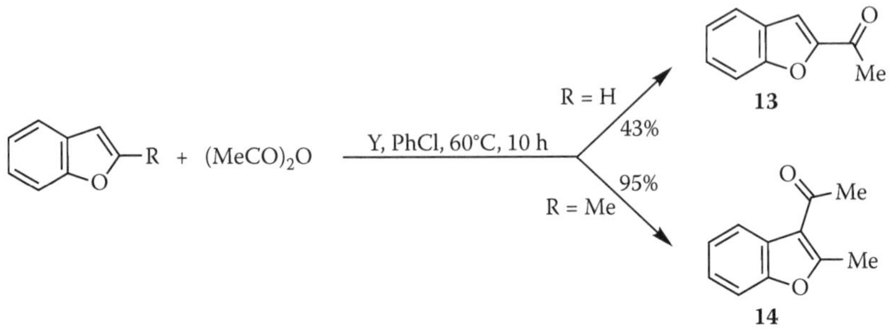

Scheme 4.8

12 h. The regenerated catalyst shows approximately the same activity as the fresh one, which indicates that the coke can be easily eliminated. In all cases, the selectivity in 2-acetylbenzofuran is between 90% and 100%.

The dealuminated Y(30) zeolite efficiently promotes the benzoylation of 2-butylbenzofuran with *para*-anisoyl chloride or *para*-anisic acid; the reaction is performed at 180°C in 1,2-dichlorobenzene (Table 4.6),[38] and leads to the three acylated isomers **15, 16,** and **17** with a good initial selectivity in the 3-acylated derivative **15.** The selectivity decreases with time due to the consecutive deacylation of this compound, followed by a reacylation that favors the formation of the other two isomers (mainly, the 6-acylated one). As expected, lower activity is observed when the carboxylic acid is utilized as acylating agent.

The same catalyst gives excellent synthetic results in the acylation of thiophene with different acyl chlorides.[39] For example, 2-butyroylthiophene is easily and quantitatively produced from the acylation of thiophene with butyroyl chloride at 100°C after 7 h. Higher homologs lead to identical results (Scheme 4.9).

Several BEA zeolites have been tested as catalysts for the liquid-phase acylation of toluene with AAN (Table 4.7). [S]-Named zeolites are hydrothermally prepared using amorphous silica as the silicon source, whereas [T]-named zeolites are prepared by using tetraethylorthosilicate (TEOS) as the silicon source with different SAR.[40] The acylation process represents a very selective route to 4-methylacetophenone **18**; small amounts of *ortho*- and *meta*-methylacetophenone **19** and **20** are produced in less than 2% of the total yield of the acetylated products. For the [S]-type BEA zeolites, the acidity values are in good agreement with the catalytic activity, following the order [S]-BEA(32) > [S]-BEA(26) > [S]-BEA(54) ≈ [S]-BEA(76). In the [T]-BEA series, the catalytic activity decreases when the SAR increases, regardless of the increase of hydrophobicity. The conclusion is that low SAR lead to better yields in acetophenones (APs). However, the poisoning

Table 4.6 Comparison of selectivity in the benzoylation of 2-butylbenzofuran with *para*-anisoyl chloride or *para*-anisic acid in the presence of Y zeolite

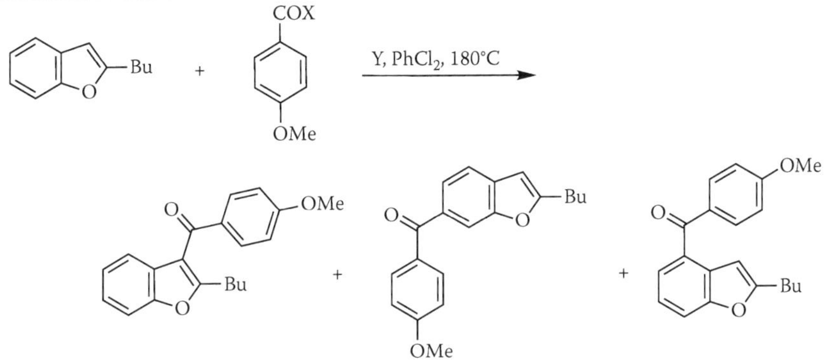

	X = Cl				X = OH			
	Conversion	Product distribution (%)			Conversion	Product distribution (%)		
t (h)	(%)	15	16	17	(%)	15	16	17
1	77	87	12	1	25	82	15	3
6	80	85	14	1	45	77	20	3
8	80	83	15	2	52	75	22	3
24	80	72	25	3	72	67	28	5

of the acid sites by adsorption of the products formed and the production of coke deposits limit the extent of the reaction.

BEA zeolites are very efficient catalysts in the benzoylation of benzene with BC.[41] The catalytic activity of BEA in comparison with ZSM-12 and REY zeolites is 5.76, 0.25, and 0.27 mmol $\times$ g^{-1} $\times$ h^{-1}. The higher activity of the BEA may be attributed to its stronger acid sites and possible contribution of the mesoporous system that minimizes catalyst decay by coke deposition, with corresponding improvements in activity and time of catalyst use. The conversion of BC to benzophenone (BP) significantly increases with increase in the reaction time, temperature, BEA:BC ratio, and benzene:BC ratio. When these parameters rise to 18 h, 80°C, 0.33 w/w ratio, and 5 mol/mol ratio, respectively, the BP yield is 54%.

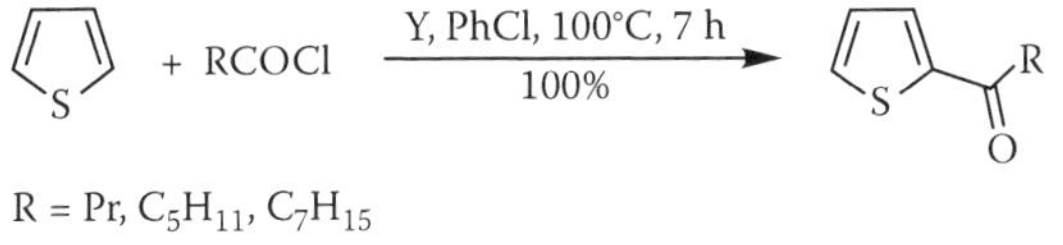

Scheme 4.9

Table 4.7 Toluene acetylation with AAN over different BEA zeolite catalysts

Catalyst	18 Yield (%)	18 Selectivity (%)
[S]-BEA(26)	54	100
[S]-BEA(32)	80	100
[S]-BEA(54)	49	100
[S]-BEA(76)	50	100
[T]-BEA(30)	54	100
[T]-BEA(100)	35	92
[T]-BEA(186)	28	89

BEA(26) also shows high activity (9.9 mmol $\times$ g^{-1} $\times$ h^{-1}) in the benzoylation of toluene with BC at 115°C. After 18 h, the conversion of BC is 83%, and the product distribution is 3, 1, and 96% of *ortho*-, *meta*-, and *para*-methylbenzophenone, respectively.[42] The catalyst can be reused after washing with acetone and calcining at 500°C. However, a slight decline in BC conversion is observed after each reuse: 83% in the first experiment, and 80% in the fourth reuse.

The BEA zeolite catalyzes the benzoylation of chlorobenzene with 4-chlorobenzoyl chloride, giving 4,4′-dichlorobenzophenone with high selectivity (~88%).[43] The conversion of the acylating agent is modest after 4 h (~20%). The presence of strong Brönsted acid sites in the zeolite catalyst is very important for the conversion of 4-chlorobenzoyl chloride into the electrophile (ClC$_6$H$_4$CO$^+$). Recycling of the catalyst still progressively decreases the acylating agent conversion to a little extent.

The same catalyst promotes the benzoylation of *ortho*-xylene with BC.[44] The conversion of BC, TOF, and selectivity for 3,4-dimethylbenzophenone are found to be 53%, 69.7 $\times$ 10^{-5} s^{-1} $\times$ mol^{-1}, and 95%, respectively, when the process is carried out at 138°C, 1 atm, and for 6 h. It is interesting to note that the non-shape-selective aluminum chloride produces higher amounts of consecutive products (16%) and gives lower selectivity for 3,4-dimethylbenzophenone (77%).

Some solid acid catalysts have been tested in the cycloacylation of 2-benzoylbenzoic acid (BBA) for producing anthraquinone (AQ).[45] In the past, the most common method for the manufacture of AQ was the

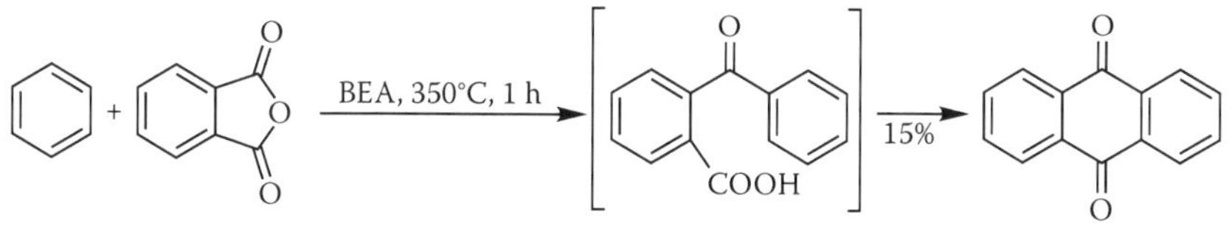

Scheme 4.10

oxidation of anthracene by different oxidants acting in the gas or liquid phase. As the availability of anthracene is declining, alternative processes have been developed and, among these, an interesting route starts from benzene and phthalic anhydride. This synthesis consists of two reaction steps: the first is usually promoted by fluoroboric acid as well as aluminum chloride; the second one, cyclization via dehydration of BBA, is commonly promoted by concentrated sulfuric acid.

Results of studies on the application of BEA in the AQ **22** synthesis by gas-phase bis-acylation of benzene with phthalic anhydride are reported (Scheme 4.10).[46] Phthalic anhydride reacts even at a relatively low temperature (<250°C), and the AQ selectivity is very high (>95%). An increase of phthalic anhydride conversion (from 6% to 16%) is observed by increasing the temperature or the time on stream (TOS), which unfortunately is accompanied by a slightly lower selectivity toward AQ, particularly for temperatures higher than 350°C. The main by-product is BBA **21**. FT-IR studies show that, at 50°C, benzene interacts with all bridged silanols and with 70% of the unbridged silanols, whereas at 250°C, these groups are completely free. Thus, at temperatures lower than 250°C, a solution of phthalic anhydride in benzene can exist inside zeolite channels, whereas at higher temperatures, benzene does not exist in the form of liquid film in the zeolite channels, and phthalic anhydride forms H-complexes with acid OH groups. At 200°C, the formation of BBA as H-complex is observed and, above 250°C, compound **21** is decarboxylated. This information allows the conclusion that AQ must be better formed at 200°C.

A further interesting application of BEA zeolite is illustrated by the direct benzoylation of arenes with BACs. The synthetic method is the subject of a patent and shows very interesting properties by the eco-efficient point of view.[47] As an example, a mixture of toluene, *para*-chlorobenzoic acid, and a little BEA (previously calcined at 400°C) pressurized with nitrogen to 2×10^5 Pa, heated to 200°C, and stirred for 4 h gives 4-chloro-4'-methylbenzophenone with 84% yield (Scheme 4.11).

Benzoylation of biphenyl (BPH) to 4-phenylbenzophenone with BC was investigated in the liquid phase over different zeolite catalysts.[48]

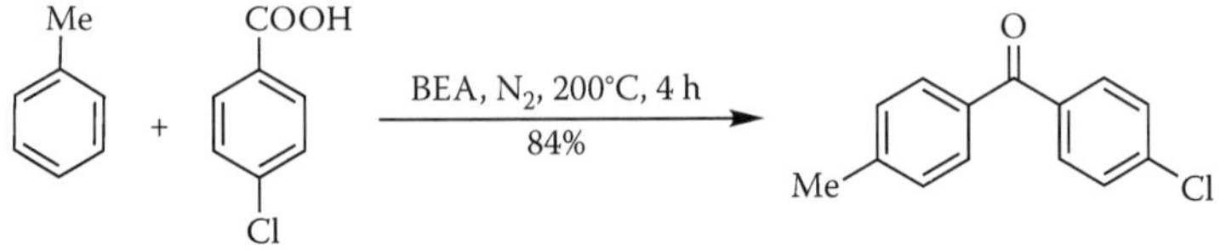

Scheme 4.11

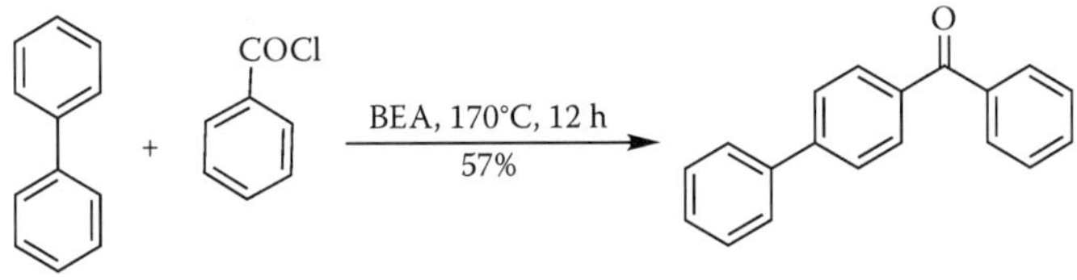

Scheme 4.12

Zeolite BEA is considerably more active than the other zeolites, affording the corresponding ketone in 57% yield and 97% para-selectivity when the reaction is carried out at 170°C for 12 h (Scheme 4.12), whereas aluminum chloride shows higher BPH conversion but less selectivity than BEA.

The conversion of BPH, TOF, and selectivity to 4-phenylbenzophenone over BEA after 6 h reaction at 170°C are 41%, 31.7×10^{-5} ($s^{-1} \times mol^{-1}$ Al), and 97%, respectively. A higher amount of 2-phenylbenzophenone is achieved over non-shape-selective aluminum chloride (20%) as compared to BEA zeolite (3%). The acidity and structure of the zeolites play an important role in the conversion of BPH and product distribution. The conversion of BPH using zeolite BEA can be significantly increased with the reaction time and BC:BPH molar ratio, being the best results achieved with 0.01 mol of BC and 0.01 mol of BPH at 170°C. The BEA zeolite can be recycled two times, and a decrease of BPH conversion is observed after each cycle, which is related to dealumination and decrease in crystallinity of the catalyst. The reaction is extended to other acylating agents, and the BPH conversion decreases in the following order: benzoylation > propanoylation > acetylation.

Acylation of BPH with AAN and hexanoic acid was also investigated (Scheme 4.13).[49] Zeolites BEA show higher activity than Y. With AAN at 83°C for 21 h by using 1,2-dichloroethane, the selectivity of the 4-acetyl-biphenyl is higher than 98%. The highest yield is ~10%, obtained with zeolites BEA(24) and BEA(100). The catalyst deactivation is mainly due to product inhibition because 4-acetylbiphenyl strongly competes for the adsorption sites on the zeolite. Acetylation of 4-acetylbiphenyl yields less than 3% of 4,4′-diacetylbiphenyl with selectivity below 12%. Acylation with hexanoic acid at 200°C without solvent affords 4-hexanoylbiphenyl in 53% yield and 92% selectivity.

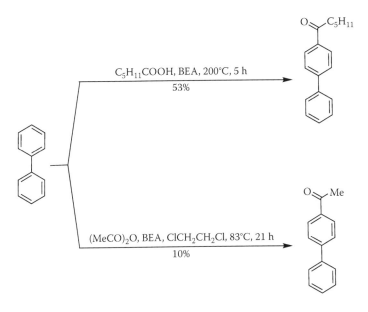

C$_5$H$_{11}$COOH, BEA, 200°C, 5 h
53%

(MeCO)$_2$O, BEA, ClCH$_2$CH$_2$Cl, 83°C, 21 h
10%

Scheme 4.13

Metal-exchanged BEA zeolites show good catalytic activity in Friedel–Crafts reactions, including acylation. The benzoylation of benzene and other aromatic compounds with BC over different Ga- and InBEA zeolites at 80°C was investigated.[50] The strong acid sites of BEA zeolite are appreciably decreased because of the deposition of gallium(III) or indium(III) oxides or indium trichloride in the zeolite, but the benzoylation activity is greatly increased. Among the modified BEA zeolites, In$_2$O$_3$/BEA shows the best performance (30% BP yield in 2 h). Concerning the benzoylation of different aromatic compounds, the activity order is benzene (30%, 2 h) < toluene (80%, 2 h) < *para*-xylene (80%, 80 min) < anisole (80%, 40 min), in agreement with a typical Friedel–Crafts electrophilic acylation mechanism. The presence of moisture in the aromatic substrate or in the catalyst has a beneficial effect, and the benzoylation rate increases appreciably. Thus, 80% conversion of BC in the benzoylation of toluene is achieved in 110 min with moist catalyst, and in 140 min with dry catalyst. The rate enhancement may be due to creation of new Brönsted acid sites by the interaction of moisture with the Lewis acid sites of the catalyst.

Metal-exchanged BEA zeolites are active in the liquid-phase acylation of toluene with AAN in nitrobenzene solvent with selectivity toward the para-isomer higher than 95% and 65% toluene conversion.[51,52] Their catalytic activity is mainly ascribed to the Brönsted acid sites. The order of activity of various metal-exchanged BEA zeolites is as follows: LaBEA > BEA ≈ CeBEA > DyBEA > EuBEA > SmBEA > GdBEA > NdBEA > PrBEA.

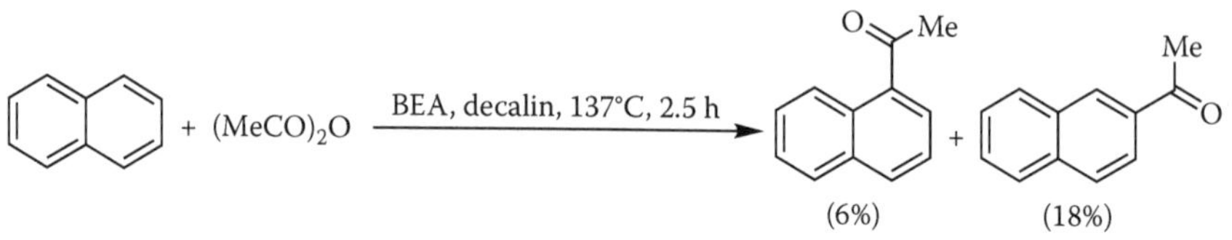

(6%)   (18%)

Scheme 4.14

BEA zeolites show good activity also in the electrophilic acetylation of naphthalene with AAN[53] and BC.[54] The BEA(75) zeolite with channel diameter 7.6 × 6.4 Å exhibits the highest activity in naphthalene acetylation with AAN (24% naphthalene conversion) (Scheme 4.14) in comparison with Y(6) with channel diameter 7.4 Å (3% naphthalene conversion), and MOR(20) with channel diameter 6.7 × 7.0 Å (3% naphthalene conversion). Reactions are performed at 137°C for 2.5 h. Despite the significant differences in naphthalene conversion observed for MOR(20) and BEA(75), practically the same selectivity to 1- and 2-acetylnaphthalene is observed (about 25% and 75%, respectively). In contrast, a lower selectivity to 2-acetylnaphthalene is observed for Y(6), which possesses large cavities in its structure that enable the formation of 1-isomer.

Similar high regioselectivity is achieved in the benzoylation of naphthalene with BC-catalyzed by BEA (Scheme 4.15).[54] 2-Benzoylnaphthalene is obtained with high selectivity (4.6/1 with respect to 1-benzoylnaphthalene), with 18% conversion of BC. A comparative study shows that BEA is more efficient than aluminum chloride and REY. Concerning the reaction mechanism, the activation of BC is still promoted by the presence of strong Brönsted acid sites in BEA that converts the BC into the electrophile PhCO$^+$.

The acylation of anisole with AAN over BEA(22) zeolite was investigated (Scheme 4.16).[55,56] The reaction is performed at 90°C for 6 h and affords *para*-methoxyacetophenone as the sole aromatic product in 95% yield. *para*-Methoxyacetophenone and AAC are produced in equal amounts at short reaction times, confirming that AAN is not noticeably hydrolyzed under the reaction conditions. Anyway, a deficit of AAC is observed at long reaction times, probably due to its reaction with the silanols of the

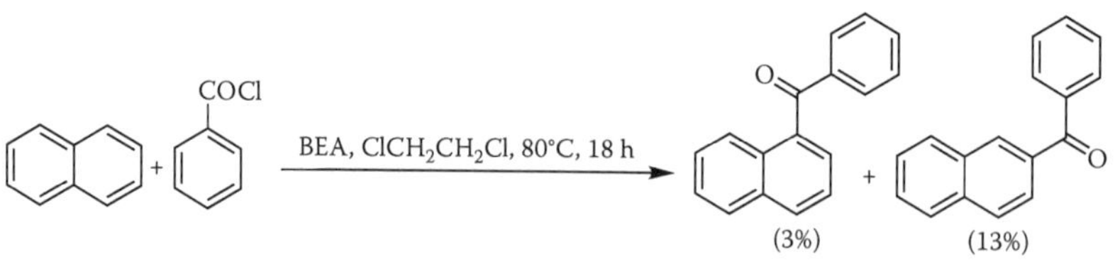

(3%)   (13%)

Scheme 4.15

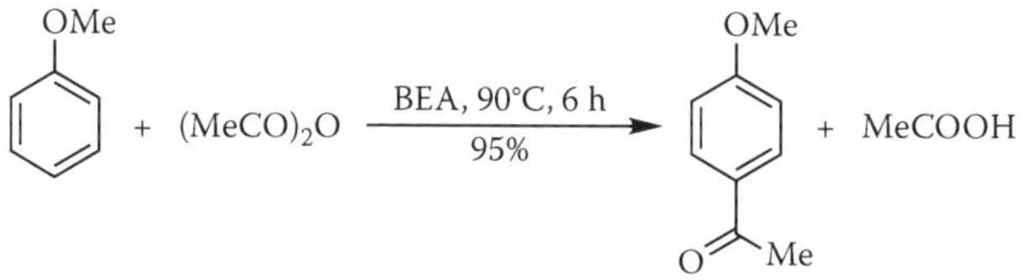

Scheme 4.16

zeolite, ultimately resulting in the dealumination of the zeolite frame-work. This uncontrollable dealumination leads to the irreversible loss of acid sites, giving rise to an irreversible catalyst deactivation.

The inhibition of the catalyst in the batch reaction was also studied by detailed kinetic analyses using a Langmuir–Hinshelwood model that allows quantification of its nature and extent. The adsorption equilibrium constant for *para*-methoxyacetophenone exceeds by a factor of at least 6 the adsorption equilibrium constant for any reactant, and the occupancy of the intracrystalline volume of the zeolite by *para*-methoxyacetophe-none increases rapidly with conversion, thereby reducing the access of reactants to the catalytic sites and resulting in catalyst deactivation as the conversion increases.[57]

The origin of the BEA deactivation in the same reaction was further investigated by different research groups.[56,58,59] By performing the reaction under batch conditions, a rapid catalyst deactivation occurs, which can be attributed, to a large extent, to the inhibiting effect of the product. However, when the reaction is performed in a fixed-bed reactor, the catalyst deacti-vation is much slower, particularly when an anisole-rich mixture is used, product *para*-methoxyacetophenone being obtained in high selectivity (>98%). The conclusion is that the use of an excess of anisole enhances cata-lyst stability and limits both retention of *para*-methoxyacetophenone and formation of polyacetylated by-products that deactivate the catalyst even if, as previously underlined, *para*-methoxyacetophenone continues to be the major poison of the catalyst. A possible acylation mechanism is proposed in which the AAN reacts first with the zeolite to form an acyl cation–zeolite complex **23** (Scheme 4.17), which then acts as acylating reagent.

Concerning the compounds adsorbed on the active sites of the cata-lyst that can decrease the acetylation rate, two classes of products have been recognized after zeolite mineralization.[60] Monoacetylated products are observed from anisole, whereas comparable amounts of mono- and diacetylated products are observed from 2-methoxynaphthalene (2-MN), *meta*-xylene, and 2-methylnaphthalene; essentially, diacetylated prod-ucts are observed from toluene. This behavior is related to the differ-ence in polarity between the substrates and the acetylated products, the

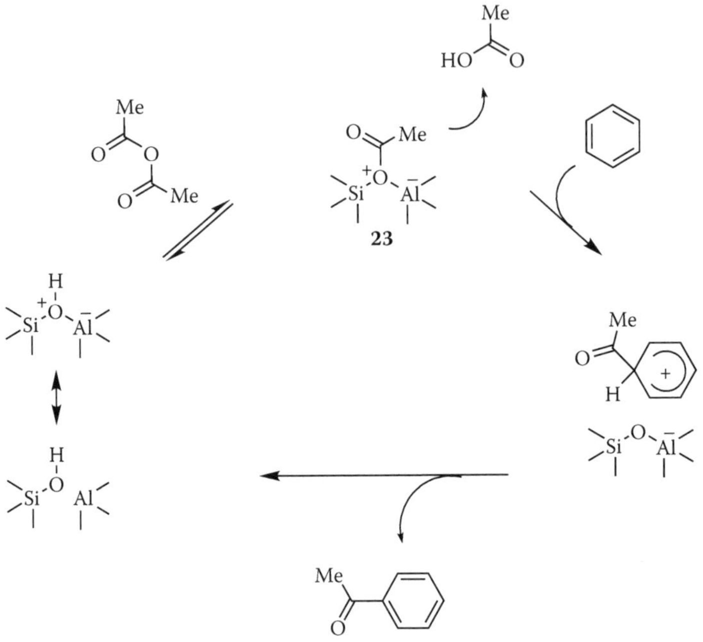

Scheme 4.17

difference between the size of substrate molecules and pore openings, and the reactivity of the substrate.

However, species resulting from AAN self-condensation, such as compounds **24, 25,** and **26,** are also present (Figure 4.5). These compounds do not appear in the acetylation of anisole and 2-MN, which shows only a limited decrease in the reaction rate with the reaction time. On the other hand, they play a more significant role than aromatic ketones in the inhibition of the acetylation reaction and are the major products retained in the zeolite during the acetylation of fluorobenzene, which occurs very slowly and with a significant decrease in the reaction rate.

Controlled dealumination is reported as an instrument to enhance the activity of zeolites BEA[61] by tailoring their acidity as we have previously shown for zeolites Y.[26] Dealuminated BEA shows, in the anisole

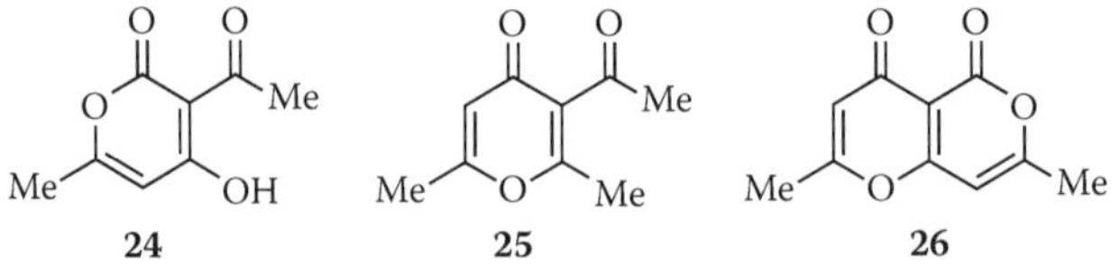

Figure 4.5 By-products derived from AAN self-condensation.

acetylation with AAN (Scheme 4.16), higher activity in the initial period compared to untreated BEA (48% versus 26% anisole conversion after 15 min), whereas at longer reaction times, the same conversion is achieved (~70% after 4 h). Dealumination provokes changes in acidity and pore distribution; the initial higher activity of the dealuminated BEA zeolite can be due to improved diffusion; this effect continues until coking reactions dominate. However, the spent catalyst's activity is restored when it is exposed to a fresh reaction mixture, which is likely able to extract the deposited higher-molecular-weight compounds, or by calcination in air at 550°C for 3 h, which removes coke by oxidation. After the preceding treatments, the catalyst can be reused in the model reaction for at least four cycles, showing quite unchanged activity (*para*-methoxyacetophenone yield: 1st = 86%, 2nd = 90%, 3rd = 87%, and 4th = 86%).[62] Therefore, two types of coke exist in the channels system and in the external surface: (1) extractable ("reversible") and (2) nonextractable ("irreversible") coke. On the basis of these conclusions, in order to increase catalyst activity, a reactor such as that shown in Figure 4.6 was utilized.[61] In this Soxhlet-like extractor-reactor, the zeolite is placed in the reflux of the condensing reaction mixture. However, the conversion is lower compared to that obtained in batch experiments (30% versus 70% after 4 h), probably due to the different vapor pressure of both reactants.

The acylation of anisole, utilized as a solvent reagent with carboxylic acids—in particular, with octanoic acid—was studied over variously activated BEA zeolites (Scheme 4.18).[63,64] Treatment of BEA with steam or acid does not result in any significant change in the surface area and pore volume. However, treating the zeolite with oxalic acid after steaming results in a significant decrease in the surface area (from 670 to 500 $m^2 \times g^{-1}$) and an increase of bulk SAR (from 13 to 51). This leads to an increase in activity and selectivity: in fact, BEA itself exhibits an activity, defined as the initial apparent first-order constant k of 0.03 $l \times g_{cat}^{-1} \times h^{-1}$ and a selectivity toward compound **27** of 80%, whereas a great activity increase (k = 0.12 $l \times g_{cat}^{-1} \times h^{-1}$) and a selectivity improvement (up to 95%) are observed with the acid-treated catalyst. It is reported that the activity enhancement observed in steamed zeolites can be due to the migration of the extraframework aluminum (EFAL) into the reaction mixture, thereby acting as an homogeneous catalyst.[13] However, the model reaction does not continue after hot filtration,[65] indicating that no active species are leached from the zeolite during the reaction. More likely, the activity and selectivity enhancement can be due to the increased accessibility or participation of active sites, rather than the formation of additional active species. Indeed, removal of EFAL and, in general, of site-blocking species results in an increase of the number of active sites that are accessible.

An efficient method for the continuous-flow selective acylation of aromatics, including ethers, over BEA zeolite, was patented.[66,67] The process

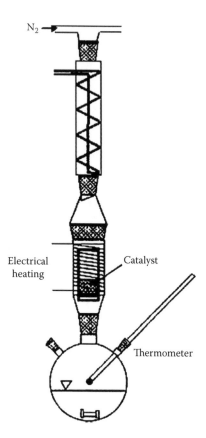

Figure 4.6 Trickle-bed reactor utilized in anisole acetylation with AAN over BEA zeolite. (From Freese, U., Heinrich, F., and Roessner, F., *Catal. Today*, 49, 237, 1999. With permission.)

is performed in a fixed-bed tubular reactor charged with BEA(25) zeolite deposited on a small layer of glass pellets or directly packed without any support or mixed with alumina. The aromatic substrate and AAN are passed through the reactor heated at 90°C–150°C for a period of 6–7 h. Selected results are reported in Table 4.8.

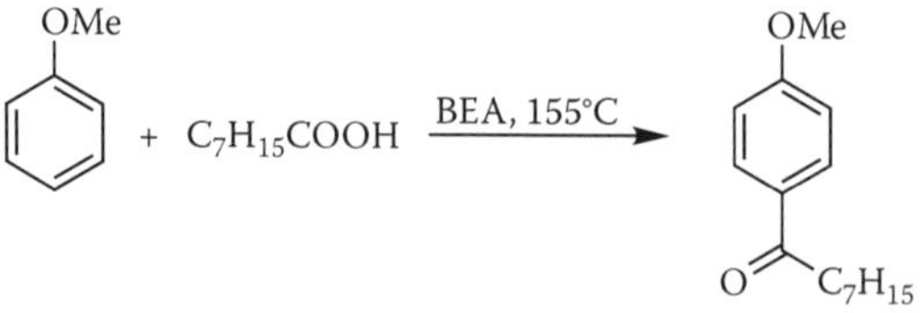

Scheme 4.18

Table 4.8 Acetylation of aromatics with AAN over
BEA in a fixed-bed reactor at 90°C–150°C for 6–7 h

Substrate	Product	Yield (%)
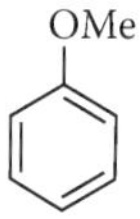 OMe	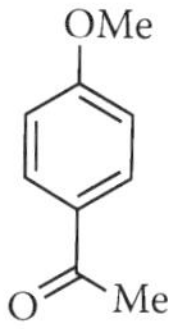 OMe ... O Me	100
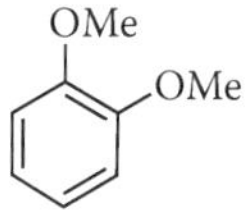 OMe OMe	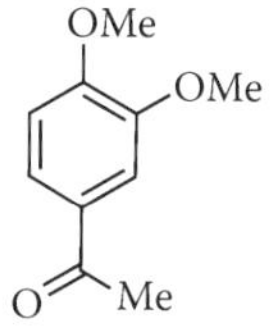 OMe OMe ... O Me	82
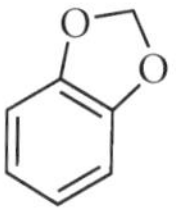 O O	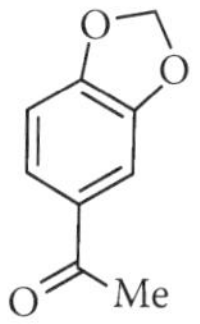 O O ... O Me	61
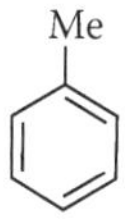 Me	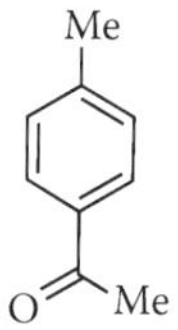 Me ... O Me	63
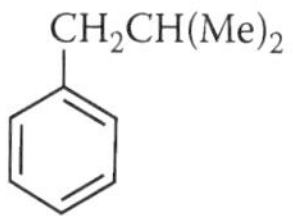 CH₂CH(Me)₂	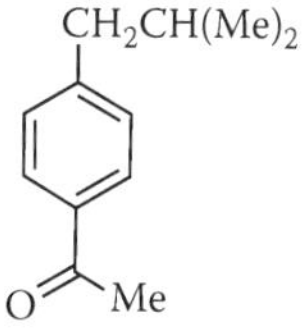 CH₂CH(Me)₂ ... O Me	75

Synthetically valuable methods for the efficient acylation of aryl ethers with anhydrides over different BEA zeolites under batch conditions are also reported.[62,68] Significant results are summarized in Table 4.9. The reaction can be performed without solvent, and the catalyst can be reused in the acetylation of anisole with unchanged activity for three further cycles after recovery and regeneration by heating in air at 550°C for 3 h.[62]

Selective propanoylation of veratrole with propanoyl chloride over BEA(26) zeolite has been performed.[69] The main product is 3,4-dimethoxypropiophenone. The performance of BEA zeolite is compared with that of

Table 4.9 Phenyl ether acylation with aliphatic anhydrides
promoted by BEA zeolite

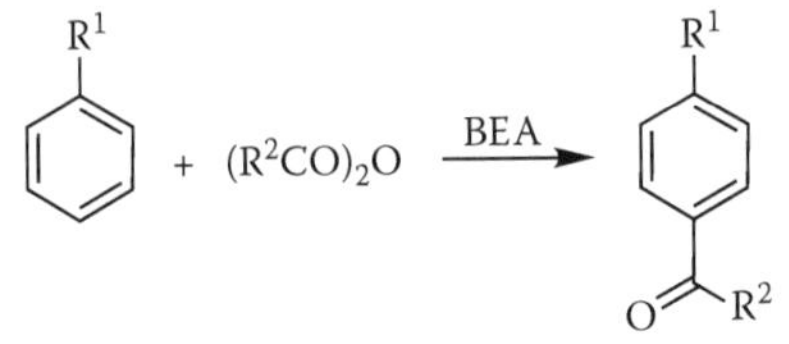

R¹	R²	T (°C)	t (h)	Yield (%)	Reference
OMe	Me	110	2.5	95	68
OMe	Et	90	3	97	68
OMe	Pr	130	3	98	68
OMe	Bu	130	3	89	68
OMe	C_5H_{11}	130	6	75	68
OEt	Me	100	3	77	62
OEt	Pr^i	100	3	75	62
OPh	Me	130	6	70	68
OPh	Et	100	3	43	62

conventional aluminum chloride. The conversion of veratrole, TOF, TON, and the selectivity to the product over BEA after 1 h reaction at 130°C are 41%, $18.8 \cdot 10^{-4}$ s^{-1} × mol^{-1}, 68, and 89%, respectively. For comparison, the conversion of veratrole, TOF, TON, and selectivity to 3,4-dimethoxypropiophenone over aluminum chloride under identical conditions are estimated to be 49%, $3.6 \cdot 10^{-4}$ s^{-1} × mol^{-1}, 1, and 69%, respectively. The catalyst can be recycled two times with a marginal decrease in catalytic activity but without loss in product selectivity.

BEA zeolite can be also applied in the acylation of different heteroaromatic compounds. Thus, a series of sulfur, nitrogen, and oxygen heterocycles can be acylated with AAN.[68] Some synthetic results are reported in Table 4.10. The best conversion and selectivity are achieved with BEA zeolite showing a surface area of 350–620 m^2 × g^{-1}.

Acetylation of thioanisole with AAN over BEA(25) zeolite can be performed, affording 4-(methylthio)acetophenone in 60% yield (Scheme 4.19).[70] The process can also be efficiently performed under continuous flow conditions. The effect of the surface acidity of the BEA zeolite catalyst is shown to play a crucial role in the regioselectivity of the reaction. In the case of BEA zeolite with 0.94 mmol × g^{-1} surface acidity, 4-(methylthio) acetophenone is obtained in 99.9% selectivity with a thioanisole conversion of 32%.[71]

Table 4.10 Heterocycle acylation with AAN in the presence of BEA zeolite

X	R¹	R²	T (°C)	t (h)	Conversion (%)	Selectivity (%)
S	H	H	80	8	70	92
S	H	H	80	3	91	70
N	H	H	80	6	40	100
N	H	H	80	2	78	64
O	H	H	25	12	26	100
O	CH=CH-CH=CH		130	6	20	100
S	CH=CH-CH=CH		130	6	30	100

The catalyst properties of and experimental conditions for the regio-selectivity of the acylation of 2-MN with AAN were carefully evaluated because different isomeric ketones can be obtained via direct acylation or isomerization (Scheme 4.20).[72-75] Concerning the special activity of BEA in the present reaction, in addition to its generally large outer surface,[76-79] it possesses particular acid properties related to local defects, that is, aluminum atoms that are not fully coordinated to the framework.[79,80] The reactions performed over this catalyst show a common behavior: initially, the kinetically favored isomer **28** is largely predominant between the reaction products; however, it disappears afterward to the benefit of isomers **29** and, in minor amount, **30**; the higher the reaction temperature, the faster this disappearance. This transformation is due to both direct isomerization of **28** into **29** and **30** and to deacetylation–transacetylation reactions in an intermolecular mechanism. From molecular modeling and analysis of the compounds entrapped in the zeolite pores during isomerization, it is possible to conclude that acetylation and isomerization mainly occur inside the zeolite pores. Compound **29** is obtained in 80% yield at relatively high temperature (≥170°C) and in the presence of a medium polar solvent such as nitrobenzene (E$_T$ 0.324).[74]

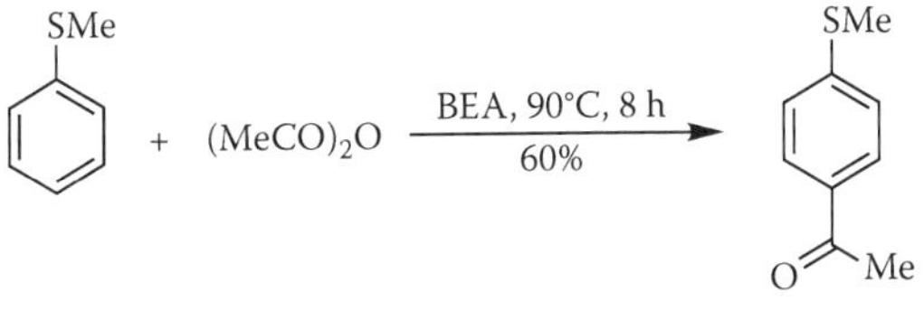

Scheme 4.19

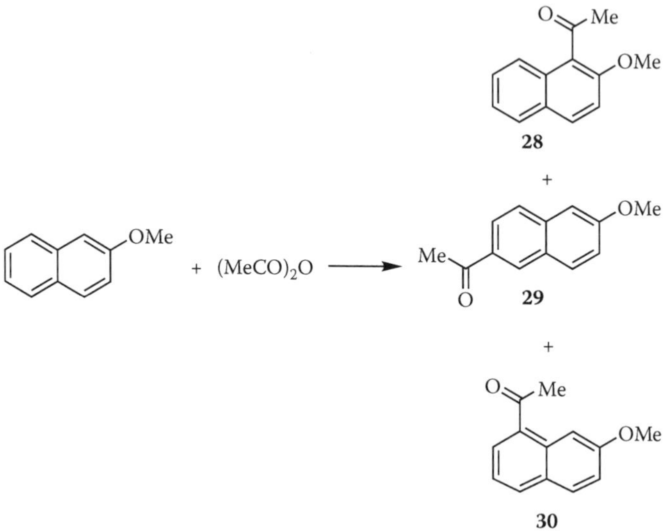

Scheme 4.20

Competitive adsorption studies of a mixture of 2-MN, product **28**, and highly polar sulfolane or 1,2 dichlorobenzene over BEA zeolite confirm the preceding conclusions and show that highly polar solvents such as sulfolane ($E_T = 0.410$, $r_0 = 184$ mmol $\cdot$ h^{-1} $\cdot$ g^{-1}) limit the entrance into the zeolite micropores of 2-MN (which is less polar) and lower the rate of the acetylation; on the contrary, solvents of intermediate polarity such as 1,2-dichlorobenzene ($E_T = 0.225$, $r_0 = 224.5$ mmol $\cdot$ h^{-1} $\cdot$ g^{-1}) favor the reaction.[74]

AC is tested in the comparative acylation of 2-MN over BEA, Y, and MOR zeolites.[81] All the three catalysts show about 30%–40% conversion of 2-MN in sulfolane in the temperature range 100°C–150°C. At 100°C, Y shows higher activity than the other two catalysts. However, as the reaction temperature is raised to 150°C, the initial high conversion drops rapidly, indicating faster deacylation of the primary product **28**. BEA and MOR zeolites also show some deacylation at higher temperatures, but the drop in conversion is not as sharp as that observed with Y. At 150°C, yields of isomer **29** are found to be similar with all three catalysts (70%–80%) due to the faster isomerization of compound **28** to the thermodynamically more favored isomer **29**.

The effect of solvent polarity was also exploited in the industrial acylation of 2-MN with a series of anhydrides (C$_2$–C$_5$).[82] In the case of AAN, the acetylation at 1-position occurs with 80% yield. The activity of different BEA zeolites passes through a maximum for an SAR framework between 60 and 80. This maximum in activity can be explained by the opposite effect of the decrease in the number of protonic sites and increase in their strength caused by dealumination and, even more significantly, by an easier desorption of the bulky acetylation products from the micropores of dealuminated samples.[73]

The influence of the acid pretreatment of BEA on its activity in the model reaction between 2-MN and AAN was analyzed in depth.[76] The contribution of the inner and outer surfaces of the catalyst was examined by considering the selectivity with respect to the bulky product **28** and the linear product **29,** which are assumed to be formed on the outer and on both the inner and outer surfaces of the catalyst, respectively.[83–85] It is shown that the production of EFAL species located in the micropores of BEA subjected to high heating rate calcination provokes the increase in the selectivity of the less hindered **29** because the formation of the bulky **28** is sterically hampered. Indeed, when the external surface of zeolite BEA is passivated by coating with amorphous silica, a significant increase in the selectivity of **29** is observed, and this result is a clear example of shape-selective acylation with zeolite catalyst. On the contrary, acid treatment increases the catalytic activity of the outer surface due to the extraction of the catalytically active EFAL species out of the micropores, leading to the preferential formation of the bulky **28**.

Quite similar good results have been achieved in the acylation of 2-MN with AAN over Y zeolite.[86] The reaction affords the more sterically hindered 1-acetyl-2-MN **28** (70% 2-MN conversion, 95% **28** selectivity). In the range of Y zeolites studied, Y(80) is shown to be the most active catalyst. The higher activity of Y(80) does not seem to be directly related to the number and/or strength of acid sites of the catalyst and, in agreement with other studies on this topic,[26] it is attributed to an increase in the hydrophobicity character of Y(80) that can favor the 2-MN approach in the zeolite framework. The conversion of 2-MN shows a level-off after 2 h that is not caused by a protiodeacylation equilibrium between 2-MN and **28**. In fact, blank experiments confirm that product **28** does not react over time and that it is strongly adsorbed onto the acid sites of the catalyst, thus acting as a poison. Deactivation of Y(80) is reversible; indeed, its catalytic activity is retained upon regeneration of the reused catalyst by thermal treatment (1st cycle: 72% conversion; 2nd cycle: 65% conversion; 3rd cycle: 72% conversion).

Comparative studies confirm the better selectivity of the zeolite USY with respect to BEA and ZSM-12 in the same acetylation reaction.[87] In fact, USY(50) produces only 1-acetyl-2-MN **28** in 60% yield and 95% selectivity, whereas lower selectivities are observed with BEA and ZSM-12 zeolites (83% and 74%, respectively).

The vapor-phase acylation of aromatic hydrocarbons with acyl chlorides over different zeolites was investigated. Comparing the acetylation of toluene with AC over ZSM-5, MOR, and REY zeolites,[88] ZSM-5 was the best catalyst affording *para*-methylacetophenone with 60% conversion of

AC and 88% selectivity; the ortho/para ratio with ZSM-5 was 0.1, whereas MOR and REY were less selective (ortho/para ratio 0.21 and 0.25, respectively). The acidic strength of ZSM-5 is stronger than that of MOR and REY: the stronger acid sites of the ZSM-5 result in higher rate of AC conversion.

The difference in the selectivity for the products over various zeolites may be interpreted in terms of geometrical constraints produced by different zeolite geometries, which have a strong impact on the reaction selectivity; geometrical constraints produced by ZSM-5 do not allow the formation of bulkier *ortho*-methylacetophenone and consecutive products in its small channels, and hence, a higher selectivity for *para*-methylacetophenone is achieved. It is also found that the conversion increases with increasing reaction temperature and toluene:AC molar ratio, whereas it decreases with the increasing reaction time, space velocity, Na-content, and SAR.

The shape selectivity showed by ZSM-5 zeolites is also exploited for the preparation of *para*-methylacetophenone in the reaction of toluene and AAN.[89] The process is carried out in batch conditions in a small autoclave with ZSM-5 zeolite previously calcined at 500°C for 5 h in a current of air. After 12 h reaction at 250°C, the para-product is obtained in 78% yield. With the regenerated catalyst (calcination in air at 500°C for 8 h), the reaction shows the same good result.

A high para-selectivity is also observed in the acylation of aromatic ethers with propanoic anhydride, carried out with the same catalyst.[90,91] The 4-methoxypropiophenone is isolated with 99% selectivity after a 2 h reaction at 150°C.

Heterocycles such as furan and pyrrole can be efficiently acetylated with AAN in the presence of zeolite ZSM-5.[92] The higher yields are 68% of 2-acetylfuran at 93% conversion of furan and 76% of 2-acetylpyrrole at 78% conversion of pyrrole over ZSM-5(30). On the contrary, the catalytic activity of Y is low due to its higher acidity compared to ZSM-5-type catalysts; this confirms that the pyrrole acylation reaction prefers moderate acidic centers.

The selectivity for acylation as well as catalytic activity are lower at high temperatures; the decrease in the yield of the acylated product at higher temperatures is due to the conversion of Brönsted acidic centers into Lewis centers.[93] The acylation proceeds largely at the 2-position of the heteroaromatics to yield a somewhat sterically bulkier product. In fact, it is not shape selectivity but electronic and thermodynamic effects that mainly determine the position of the acylation in these substrates.

For the gas-phase acylation reaction of aromatics with anhydrides, metal-exchanged zeolites are also employed. In the presence of CeZSM-5, the acylation of benzene with AAN affords AP with 86% conversion of AAN and 95% selectivity;[94] on the contrary, unmodified ZSM-5 is found to be less active under the same reaction conditions (71% conversion of AAN, 86% selectivity to AP). If the reaction is carried out with AAC as acylating

Table 4.11 Comparison of the activity and selectivity of ZSM-5 and MOR zeolites in the gas-phase acylation of benzene with AAC at 250°C

Catalyst	TOF (S⁻¹)	AAC conversion (%)	Product distribution (wt%)	
			AP	Others
ZSM-5(41)	46.2	43	91	9
MOR(22)	4.3	7	31	69

agent, a conversion decrease is observed (from 85% to 48%) while maintaining a good level of selectivity (87%).

Carboxylic acids are employed as acylating agents in the gas-phase acylation of aromatic substrates with zeolite. Acetophenone is prepared starting from benzene and AAC in a flow reactor under atmospheric pressure in the presence of catalysts.[95] The zeolite catalyst is packed in a glass reactor and placed in a vertical furnace, and a mixture of benzene and AAC is passed through the catalytic bed by a syringe pump. The results of the catalytic activity using ZSM-5 and MOR catalysts are shown in Table 4.11. The main product of the reaction carried out with ZSM-5 is AP; small amounts of diacetylbenzenes are also detected. After the initial 2 h run, ZSM-5 exhibits higher catalytic activity, and the TOF is found to be several times higher over ZSM-5 than MOR (46 m mol s⁻¹ mol⁻¹ Al × 10⁻³ and 4 m mol s⁻¹ mol⁻¹ Al × 10⁻³, respectively). Zeolites Y and REY are ineffective for the acetylation of benzene to AP under the preceding conditions. The conversion of AAC over ZSM-5 and MOR is 43% and 7%, respectively. In addition, the selectivity for AP over ZSM-5 is greater (91%) than that of MOR (31%). From these data, it is clear that the performance of the catalyst in this case is governed by the acidity and structure of the zeolite. The higher conversion of AAC over ZSM-5 seems to be due to the fact that ZSM-5 exhibits a three-dimensional channel system and stronger Brönsted acid sites compared to MOR. The acid strength, in other words, is the crucial factor for the conversion of AAC into the electrophile ($MeCO^+$), which attacks the benzene ring and produces AP. Moreover, the higher selectivity for AP over ZSM-5 can be attributed to the smaller pore openings (0.51 × 0.55 nm and 0.54 × 0.56 nm) than those of MOR (0.65 × 0.7 nm), which prevent the secondary reactions in the zeolite channels and, consequently, the formation of diacetylbenzenes.

NaZSM-5 zeolites are less active but more selective than ZSM-5. The AAC conversion rate decreases in the following order with respect to sodium content: ZSM-5 > Na(29)ZSM-5 > Na(34)ZSM-5 > Na(37)ZSM-5. These data suggest that the stronger Brönsted acid sites of ZSM-5 are capable of facilitating reaction between AP and AAC (to a small extent) and, consequently, the formation of diacetylbenzenes. Thus, removal of a small amount of stronger Brönsted acid sites by sodium exchange in

Table 4.12 Anisole acylation with different carboxylic acids catalyzed by zeolite ZSM-5

			Product distribution (%)	
R	T (°C)	Conversion (%)	31	32
Me	120	5	0.5	99.5
Et	147	92	60.0	40.0
Pr	150	73	83.6	16.4
Bu	150	38	82.1	17.9
C_5H_{11}	150	24	83.0	17.0
C_6H_{13}	150	9	81.2	18.8
C_7H_{15}	150	5	80.0	20.0
$C_{11}H_{23}$	150	1	—	—
$C_{17}H_{35}$	150	1	—	—
Ph	150	0	—	—

ZSM-5 results in slightly lower activity accompanied by higher selectivity. In addition, the temperature of the reaction has a determining influence on the rate of AAC conversion and product distribution. The TOF over ZSM-5 linearly increases with the increase in reaction temperature, but selectivity toward AP decreases: a maximum in AP selectivity (100%) over this catalyst is observed at 230°C, but the rate of AAC conversion is found to be very low (TOF = 16 m mol s^{-1} mol^{-1} Al $\times$ 10^{-3}).

The process can be carried out with aromatic substrates such as benzene, toluene, and isopropylbenzene and different acylating agents (56%–99% conversion).[96] Zeolite ZSM-5 fails to give any product with other substrates such as xylenes, mesitylene, and *N,N*-dimethylaniline.

Anisole can be acylated with carboxylic acids over ZSM-5 catalyst.[97] The reaction is carried out in a batch reactor under reflux for 2 d. Table 4.12 shows the results obtained utilizing carboxylic acids with different chain length. From propanoic acid to stearic acid, the conversion decreases from 92% to 1%, showing a trend opposite to that observed with REY zeolites.[10] This behavior can be ascribed to the small micropore size of ZSM-5, in which the formation of large molecules is difficult (for pentanoic to octanoic acid) or impossible (for longer-chain carboxylic acids). The poor conversion of AAC can be attributable to the low reflux temperature.

From a theoretical point of view, the products more easily produced in the acylation of anisole are the *ortho-* and *para*-acylanisoles and phenyl carboxylic ester **32**.[63,64] The selectivity of ZSM-5 zeolite toward *para*-acylanisole **31** increases from 0.5 with AAC to 60% with propanoic acid and finally reaches 80% with butanoic acid. In addition, at low temperature (120°C), the reaction affords product **32** with a selectivity higher than 70%, whereas the maximum selectivity toward ketone **31** is achieved by carrying out the process at 150°C. This means that the activation energy of the O-acylation reaction is lower than that of C-acylation.

The handling of zeolites in batch reactions frequently shows some drawbacks since these materials have a colloidal form, and the small size of the particles renders difficult their recovery and purification. The transformation of powders into extrudates or tablets requires the introduction of binders such as α-alumina that can increase the diffusion barrier between the reactants and the active sites. For these reasons, zeolites have been supported on special materials to obtain a macroscopic shaping easily recoverable from the reaction mixture and more utilizable in a fixed-bed reactor.[98,99]

Zeolite BEA(75) coatings can be deposited on a ceramic cordierite $(2Al_2O_3 \cdot 5SiO_2 \cdot 2MgO)$ monolithic substrate. The coating is performed with a suspension of zeolite, with silica as the binder, and a surfactant in water. The mixture is well stirred; the cordierite monolith is dipped into the mixture and dried at 200°C, and finally is calcined at 400°C for 4 h.[64]

The BEA-coated monolith exhibits an activity of $0.034 \ 1 \times g^{-1} \times h^{-1}$ in the acylation of anisole with octanoic acid (Scheme 4.18). The activity is lower than in a slurry reactor $(0.055 \ 1 \times g^{-1} \times h^{-1})$ in which the catalyst surface is optimally used. Both the slurry and monolithic approaches show good selectivities (78% for monolith and 86% for slurry). The coated monolith is tested in a larger scale (2 m reactor length). The reactor can be utilized for two runs, giving similar conversions (80% in 8 h at 155°C). The use of monolithic reactors allows easy removal of the produced water by a countercurrent stripping operation using a gas flow through the reactor.

The preparation and use of active catalysts coated on a structure packing was further studied as an attractive replacement for conventional catalysts in randomly packed beds or slurry reactions.[100] A method was developed in which catalytically active and selective BEA coatings could be prepared onto ceramic monoliths constituted either of pure silica or cordierite (Figure 4.7a) and metallic wire gauze packings (Figure 4.7b).[101]

The average loading values of the BEA coatings are the following: silica monolith, 6.6%; cordierite monolith, 5.4%; and silica-precoated wire gauze packing, 5.6 wt%. The activity of the coated structures is measured batchwise in the acylation of anisole utilized as a solvent–reagent with octanoic acid. The liquid reaction mixture is carried through the monoliths using a nitrogen flow. The silica monolith shows the highest activity

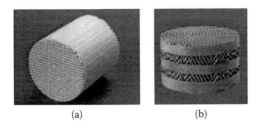

Figure 4.7 Ceramic monolithic supports for BEA zeolite utilized in the acylation of anisole with octanoic acid. (From Beers, A. E. W., Nijhuis, T. A., Aalders, N., Kapteijn, F., and Moulijn, J. A., *Appl. Catal. A: Gen.*, 243, 237, 2003. With permission.)

(k = 0.050 l × g_{cat}^{-1} × h^{-1}), which is only 10% lower than that of BEA particles in slurry. The activity of the coated cordierite monolith and the metal wire gauze packing is much lower than that of the BEA particles in slurry. All monoliths give a similar selectivity of about 77%–85% toward the main para-isomer **27** compared to the 84% selectivity of the original BEA tested in slurry. The BEA-coated monoliths can be reused after regeneration by washing with acetone, drying at 120°C (16 h), and calcining at 450°C for 4 h.

Benzoylation of anisole with BC (Scheme 4.21) was investigated over BEA zeolite, supported on macroscopic preshaped SiC material (BEA/SiC) as a high-performance catalyst.[102] The SiC support exhibits high thermal conductivity, high resistance toward oxidation, high mechanical strength, and chemical inertness properties required for good support. The BEA zeolite is synthesized and in situ deposited on SiC extrudates previously prepared by a reported technology.[103] A high-magnification SEM image shows the presence of small zeolite particles homogenously dispersed on the entire surface of the support. The deposition of BEA on the SiC surface significantly increases the overall surface area of the material from 25 (for pure SiC) to more than 100 m^2 × g^{-1} (for BEA/SiC), along with a large microporous contribution (~60 m^2 × g^{-1}). This composite catalyst shows relatively high activity with a total 71% conversion of BC after 24 h on stream (95% selectivity with respect to *para*-benzoylanisole and less than 5% selectivity for *ortho*-benzoylanisole). The catalyst can be reused after

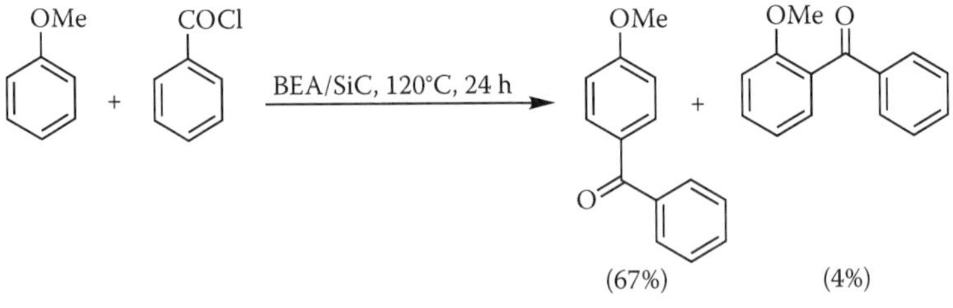

OMe   COCl   OMe   OMe O

      + $\xrightarrow{\text{BEA/SiC, 120°C, 24 h}}$ +

(67%)   (4%)

Scheme 4.21

washing with methylene chloride, showing quite similar results for three tests. Similar experiments carried out with unsupported higher-acidic BEA show that, in the first cycle, similar activity and ketone selectivity are achieved, but significant catalyst deactivation is observed during the second cycle. The higher activity of BEA/SiC is probably due to the dispersion of the zeolite on the support surface, leading to a higher contact surface with reactants.

With the aim of achieving a supported BEA with higher performance, BEA zeolite nanowire material was prepared using mesoporous multi-walled carbon nanotubes as directing templates.[104] The catalyst can be obtained by growing the BEA zeolite in the presence of carbon nanotubes favoring the formation of carbon templates. It is assumed that the confinement effect inside the tubes can induce an increase in pressure, thus promoting the synthesis of nanoparticles of BEA zeolite by replacing the hydrothermal pressure increase by the internal pressure increase. The final carbon template is removed by submitting the sample to calcination in air at 750°C. The as-prepared BEA zeolite is then converted into its H-form, and it is employed in the acylation of anisole with BC (Scheme 4.21), showing satisfactory activity (anisole conversion ~65%) and selectivity toward *para*-methoxybenzophenone (~95%). However, it must be emphasized that similar and even better results can be obtained in the same reaction promoted by simpler and less expensive catalysts.

The use of Y, BEA, and ZSM-5 zeolites in the Friedel–Crafts acylation of aromatics shows some limitations when large-sized molecules are utilized. In fact, because of the dimension of their pores, the access to the internal active sites of these zeolites is restricted to molecules with kinetic diameter up to 8 Å. Trying to overcome this drawback, mesoporous molecular sieve MCM-41 can be utilized as interesting catalyst or support for catalysts for reactions involving larger molecules.

The preparation and use of indium trichloride, gallium trichloride, and zinc chloride supported on MCM-41 as Lewis acids in the Friedel–Crafts acylation of aromatics with acyl chlorides was investigated.[105,106] The support itself shows no catalytic activity in the benzoylation of benzene with BC, whereas the highest activity is showed by the supported indium trichloride. The order for acylation activity of the supported metal chloride (indium trichloride > gallium trichloride >> zinc chloride) is quite similar to that of the redox potential of the metals $[E^0_{In^{3+}/In}$ (−0.34 V) > $E^0_{Ga^{3+}/Ga}$ (−0.53 V) > $E^0_{Zn^{2+}/Zn}$ (−0.74 V)] and confirms a possible relationship between the redox potential and the catalytic activity of the supported metal chloride.[107] The reaction can be efficiently applied to a variety of aromatic compounds, including toluene, *para*-xylene, mesitylene, anisole, and 2-MN (70%–90% yield), confirming the moisture insensitivity of the catalyst.[106]

More interesting are the results on the support of indium oxide, gallium oxide, and zinc oxide on MCM-41.[108,109] In particular, In$_2$O$_3$/MCM-41 proved to be a highly active catalyst for the acylation of aromatic compounds with BC even in the presence of moisture. The activity order of the three supported metal oxides is found to be the same as that for the supported metal halides.[109] The times requested for half of the benzene benzoylation in different solvents in the presence of In$_2$O$_3$/MCM 41 are in the following order: methylene chloride (2.8 h) < acetonitrile (3.2 h) < heptane (6.2 h). These results indicate that the catalyst is more active in the presence of polar solvents even if a parallel correlation with Reichardt E_T^N polarity parameters[110] (methylene chloride [0.309], acetonitrile [0.460], and heptane [0.012]) is not observed.

Aluminum-chloride-grafted MCM-41 shows high catalytic activity in the benzoylation of benzene and toluene with BC,[111] leading to 55% and 63% BC conversion, respectively. The catalyst can be reused in the subsequent experiments but with somewhat reduced activity; this is most probably due to its momentary exposure to the atmosphere during its reuse. When the catalyst is exposed to the atmosphere for a longer period (>2–3 min), its activity is totally killed and, also, its color changes from brown to white due to the adsorption of moisture from the atmosphere.

The mesoporous aluminosilicate AlMCM-41-type material, prepared from BEA zeolite seeds, can be utilized as catalyst in the acylation of anisole with acyl chloride with the aim of improving the transport of the reactants, especially for the relatively hindered molecules such as octanoyl chloride.[112] The catalyst shows good activity, being *para*-octanoyl anisole obtained in 90% yield after 1 h. As commonly observed, the use of carboxylic acid as acylating agent results in a slower process (~20% yield after 26 h) due to its lower electrophilicity and the production of water that inhibits the active sites of the zeolite, as previously observed by Beers et al.[113]

The acylation of 2-MN with different anhydrides at low temperature (100°C) in the presence of the moderately protic acid MCM-41 mesoporous zeolite was studied.[114] The selectivity toward 1-isomer **28** is practically 100% with the less bulky AAN (Scheme 4.20), whereas with the more bulky isobutyric anhydride, a small amount of 6-isomer is produced. At a higher temperature (~130°C), the reversibility of the acylation at the 1-position plays a fundamental role in the distribution of products, and a decrease in the selectivity for 1-position is observed. The catalyst can be recycled three times; after each cycle, the catalyst is filtered, washed, dried, and reactivated by calcination for 2 d at 450°C, showing quite similar conversion (~40%) and selectivity (~97%) with respect to compound **28**.

4.2 Clays

Clay minerals occur abundantly in nature, and their high surface area, and sorptive and ion-exchange properties, have been exploited for catalytic applications through decades.[115,116] Solid clay materials[117] have a broad range of functions in catalysis, including the use as (1) catalytically active agents (usually as solid acids), (2) bifunctional or "inert" supports, and (3) fillers to give solid catalysts with required physical properties.

Clay minerals are constituted of layered silicates. They are crystalline materials of very fine particle size ranging from 150 to less than 1 μm. There are two basic building blocks, such as tetrahedral and octahedral layers, which are common to clay minerals.[118] Tetrahedral layers consist of continuous sheets of silica tetrahedral linked with three corners to form a hexagonal mesh and the fourth corner of each tetrahedra in adjacent layers. Octahedral layers, on the other hand, consist of flat layers of edge-sharing octahedral, each formally containing cations at its center (usually magnesium or aluminum) and hydroxy and oxy anions at its apices.

The different classes of clay minerals, namely, 1/1, 2/1, and so on, have a different arrangement of tetrahedral and octahedral layers. Structural units of clays, therefore, consist of either (1) alternating tetrahedral sheets (OT or 1/1 structure; e.g., the kaolinite group); (2) a sandwich of one octahedral sheet between two tetrahedral sheets (TOT or 2/1 structure; e.g., smectite clay minerals, of which the most common member is montmorillonite); or (3) an arrangement in which three TOT units alternate with a brucite layer (2/1/1 structure; e.g., chlorite).

The smectite group mineral is one of the most abundant. This group of clay minerals has a dioctahedral or trioctahedral 2/1 layer structure with isomorphous substitution that leads to a negative layer charge of less than 1.2 per formula unit. Interlayer spacings vary between ~10 and 15 Å and are generally dependent on the nature of the exchangeable cations and the relative humidity. Smectites are divided into four subclasses, depending upon (1) the type of octahedral layer (dioctahedral or trioctahedral) and (2) the predominant location of layer charge sites (octahedral or tetrahedral).

Clay minerals that predominantly have properties governed by smectite are called *bentonite*. Montmorillonite (Figure 4.8) is a major constituent of most bentonites (typically, 80%–90%), the remainder being a mixture of mineral impurities, including quartz, crystobalite, feldspar, and various other clay minerals, depending upon geological origin. Bentonites containing high levels of montmorillonites are the most abundant and commercially available forms of smectite.

Montmorillonite shows a high cation exchange capacity (CEC) that allows a wide variety of catalytically active forms of the clay mineral to be

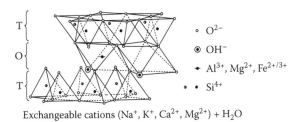

T { O²⁻
O { ◉ OH⁻
T { ⁺ Al³⁺, Mg²⁺, Fe²⁺/³⁺
     ∙ ∙ Si⁴⁺

Exchangeable cations (Na⁺, K⁺, Ca²⁺, Mg²⁺) + H₂O

Figure 4.8 Montmorillonite structure. (From http://www.scielo.br/. With permission.)

prepared (e.g., containing acidic cations, metal complexes, photocatalytically active cations).

Montmorillonites are more frequently used as Brönsted acid catalysts even if Lewis acidity plays a role in their catalytic activity. The origin of Brönsted acidity in metal-exchanged montmorillonites is ascribed to the polarizing influence of the cation on the water molecules in spatially restricted interlayers.[119] The exchangeable cations are either protons or polarizing cations [e.g., aluminum, chromium(III), or iron(III)].

Acid site strength depends upon the type of interlayer cations present ($H_3O^+ > Al^{3+} > Ca^{2+} > Na^+$). Higher acid strength generally leads to greater catalytic activity but poorer product selectivity. Controlling the acid site strength by choice of interlayer cations proves to be useful for fine-tuning the catalyst selectivity. When treated with mineral acids under harsh conditions (e.g., refluxing), montmorillonites undergo leaching of aluminum and, to a lesser extent, silicon. This leads to increased surface area and concentration of weak acid sites accompanied by a decrease in the concentration of strong acid sites. Acid-leached montmorillonites are particularly useful for catalytic applications requiring only weak acid sites because strong acid sites give rise to poor selectivity.

Isomorphous substitution of cations in the lattice by lower-valent ions—for example, substitution of aluminum for silicon, magnesium or iron(II) for aluminum, or sometimes lithium for magnesium—leaves a residual negative charge in the lattice that is usually balanced by sodium ions.[118] These can be readily replaced by other cations when brought into contact with these ions in aqueous solution with the goal of preparing specific heterogeneous metal catalysts.

Swelling is a further important property of clays. In fact, many clay minerals adsorb water between their layers, which move apart and the clay swells. For efficient swelling, the energy released by cations and/or layer solvation must be sufficient to overcome the attractive forces (such as hydrogen bonding) between the adjacent layers. In 2/1 (TOT) clay minerals such as smectite, during swelling, the interlayer cations can undergo exchange with external solutions. Since smectites have the highest concentration of interlayer cations, they have the highest CEC (70–120 meq × 100 g^{-1}).

The interlayer cations contribute to the acidity of clay minerals. Some of these cations may be protons or polarized cations (e.g., aluminum) that give rise to strong Brönsted acidity.[120] The higher the electronegativity of the cation, the stronger are the acidic sites generated. Brönsted acidity also stems from the terminal hydroxy groups. In addition, clay minerals have layer surface and edge defects that would result in weaker Brönsted and/or Lewis acidity, generally at low concentrations. The acidity of clay minerals can be comparable to that of concentrated sulfuric acid. The surface acidity (Hammet function H_0) of natural clays with sodium or ammonium ions as interstitial cations ranges from +1.5 to −3. Washing with mineral acid, such as hydrogen chloride, brings down the H_0 value from −6 to −8, which is between concentrated nitric acid (−5) and sulfuric acid (−12).

Finally, between the TOT layers of a smectite, large cationic species that are polymeric or oligomeric hydroxyl metal cations formed by the hydrolysis of metal salts of aluminum, gallium, chromium(III), silicon, titanium(IV), iron(III), and mixtures of them can be inserted by cation exchange, giving the so-called pillared clays.[121]

A second class of clays, namely, the hydrotalcite-like anionic clays, represents a great number of layered materials that are synthetic or natural crystalline products consisting of positively charged two-dimensional sheets with water and exchangeable charge-compensating anions in the interlayer region.[122,123] Hydrotalcites are anionic clays and are less diffuse in nature than cationic; they can be used as such or (mainly) after physicochemical treatment. The most interesting properties of the resulting materials are the high surface area, basic behavior, and possibility to form uniform mixtures of oxides with very small crystal size stable to thermal treatment. These properties have found application in the field of heterogeneous catalysis where hydrotalcites are directly utilized as catalysts or as support for catalysts.

Unmodified acid clays have been little utilized as catalysts for Friedel–Crafts acylation reaction. Among the scanty information from the literature, the production of AQ **22** through cyclodehydration of BBA **21** is of some interest (Scheme 4.22).[124] The reaction is performed in the presence of different clays at 350°C for 30 min. AQ **22** is directly recovered by sublimation from the reaction mixture. Some interesting results are achieved with commercially available catalysts such as Tonsil BW3, a mixture of mica, quartz, feldspar, and amorphous silica with bentonite and montmorillonite K10, product **22** being obtained in 78%–89% yield. Reusability studies show that Tonsil BW3 still gives product **22** in 60% yield after five

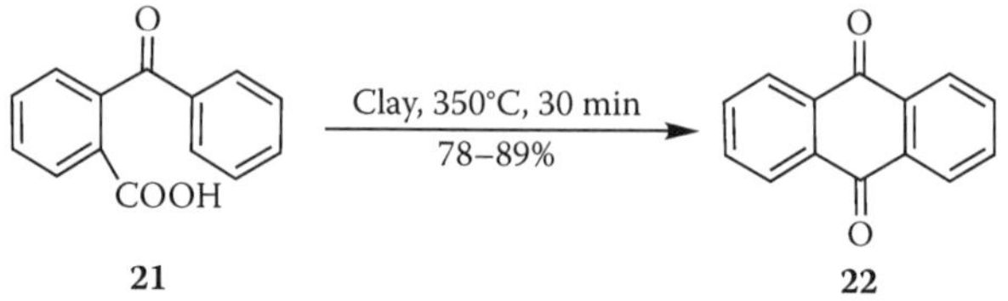

Scheme 4.22

cycles. The catalyst deactivation is ascribed to the progressive piling up of product **22** into the catalyst surface.

The same reaction can be performed by using microwave energy instead of conventional heating.[125] In this case, a mixture of **21** and bentonite is irradiated in a microwave oven at a power of 600 W for 5 min, raising the temperature from 350°C to 400°C. The AQ is still recovered by sublimation, and the catalyst can be reused for 15 cycles, giving product **22** with >90% yield.

Heterocyclic quinones **34** are similarly prepared in 21%–92% yield from 2-(2-thiophenecarbonyl)benzoic acids **33** with montmorillonite K10 purified by removal of quartz and feldspar, and using microwave irradiation at a power of 780 W for 3–5 min as an energy source (Table 4.13).[126]

Clay-based materials prepared by metal exchange or Lewis acid impregnation have received great attention as catalysts in liquid-phase organic reactions, including Friedel–Crafts acylations.[115,116] Clayzic is zinc chloride impregnated on acid-treated montmorillonite K10[127]; spectroscopic analysis shows that, in clayzic, the active sites are largely Lewis acid in character. Thus, clayzic owes its remarkable activity in Friedel–Crafts

Table 4.13 Cycloacylation of 2-(2-thiophenecarbonyl)benzoic acids promoted by montmorillonite K10

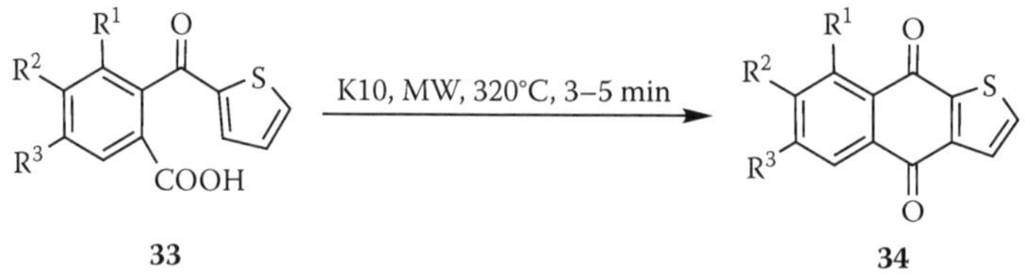

R^1	R^2	R^3	Yield (%)
H	H	H	60
Cl	Cl	H	92
H	NO_2	H	21
H	Me	Me	27
H	H	OMe	43

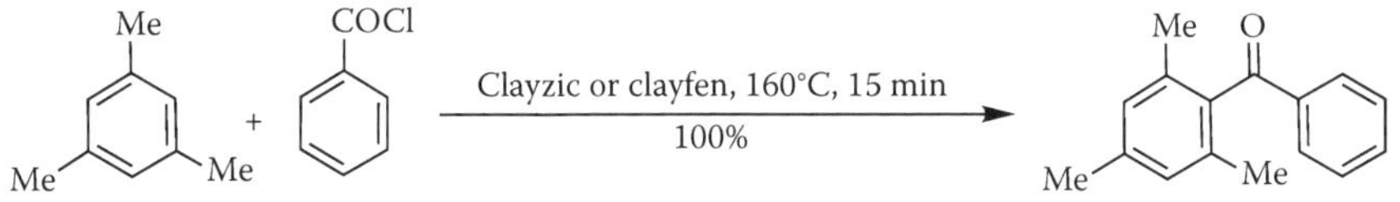

Scheme 4.23

reactions to the presence of high local concentration of zinc ions in structural mesopores.[127] Clayzic can be applied as a solid catalyst in the acylation of aromatic compounds with acyl chlorides and anhydrides. Comparison of catalytic activity with various metal-exchanged K10 montmorillonites was made.[128] In the benzoylation of mesitylene with BC at 160°C, nearly a quantitative yield of 2,4,6-trimethylbenzophenone is achieved in about 15 min with both clayzic as well as K10 montmorillonite exchanged with iron(III) cations ("clayfen") (Scheme 4.23). Good results are also achieved with other metal-exchanged K10 clays under the same conditions (AlK10: 98%; CrK10: 94%; TiK10: 100%; CoK10: 84%).

In a comparative study on the acylation of mesitylene and anisole with BC, promoted by clayfen,[129–131] it is observed that mesitylene reacts faster (by a factor of three) when the reactions are carried out separately. However, when the two substrates are reacted jointly, competition favors anisole (by the same factor of three). A tentative explanation for this reversal activity can be based on the assumption that benzoylation of mesitylene and anisole follows two different mechanisms characterized by different levels of interaction of the catalyst surface with the aromatic substrate and the benzoyl cation. In particular, it must be taken into consideration that anisole interacts more strongly than mesitylene with the acid centers of the catalyst and, consequently, its intrinsic higher reactivity toward the acyl cation can be dramatically lowered, and the acid sites are less available for mesitylene acylation reaction.

Acylation of toluene with carboxylic acids can be performed with metal–clay-exchanged catalysts.[132,133] A series of exchanged montmorillonites was studied in the model acylation of toluene with dodecanoic acid. Best results are achieved with AlK10 (60% yield of the three isomers).

Thanks to their low cost and easy availability, many metal-exchanged clays have been patented as efficient solid catalysts in Friedel–Crafts acylation reactions. A great number of arylketones is prepared by electrophilic acylation of arenes with anhydrides in the presence of ion-exchanged clays at 150°C–250°C.[134] Thus, for example, aluminum-enriched mica promotes the reaction of BAN with *meta*-xylene at reflux for 4 h in 99% yield.

BACs can be activated toward Friedel–Crafts acylation with commercially available catalyst Envirocat EPIC®.[135] The catalyst is constituted by a natural clay impregnated with polyphosphoric acid. The reaction occurs satisfactorily only with anisole and with variously substituted BACs, giving the corresponding BPs in 20%–69% yield.

The more reactive aromatic ethers are acylated with anhydrides over clayzic or clayfen.[136] 2-MN selectively gives 1-acetyl derivative **28** in 71% yield at 80°C in nitrobenzene, whereas at 140°C the yield increases until 82% but a drop in selectivity is observed, small amounts of thermodynamically favored 6-acylated product **29** (~7%) accompanied by the 8-acylated product **30** (~2%) being obtained via protiodeacylation (Scheme 4.20).

Highly selective acylation at the 4-position is achieved by the reaction of propanoic anhydride with anisole (68% yield) and veratrole (52% yield) in the presence of clayzic at 90°C for 10 h in nitrobenzene. The catalytic activity depends on the proper balance of Brönsted and Lewis acid sites present in the clayzic.

Rare-earth-exchanged clays were tested in the benzoylation of anisole with BC.[137] The optimum activity (93% anisole conversion after 20 h) is obtained using LaK10 heated at 280°C. It is shown that, while lanthanum trichloride is the predominant species in the catalyst structure at 280°C, a more complex mixture is formed at a higher temperature (i.e., 550°C and higher), including lanthanum(III) oxychloride and oxide, but the presence of species linked to the support by oxygen–silicon bonds is also possible; therefore, the support also plays a role in the catalytic activity. It is suggested that, during the reaction, chloride species can be regenerated from the supported species by reaction with hydrogen chloride produced from acyl chloride. When lanthanum chloride is impregnated on silica and heated at 280°C, lower reaction rates are observed (70% anisole conversion after 20 h). Moreover, the heating temperature (from 280°C to 950°C) does not strongly affect catalyst performance. The reaction with LaK10 can be extended to different acyl chlorides and aromatic and heteroaromatic substrates, giving arylketones with acylation rates up to 2000 cycles × h^{-1}.

Heterocyclic compounds, namely, pyrrole, thiophene, and furan, can be efficiently acylated with AAN in the presence of different metal-exchanged K10 clays, in particular, with clayfen (Scheme 4.24).[138] 2-Acetylheteroaromatics **35** are produced with high heterocyclic conversions (97%–98%) accompanied by excellent selectivities (92%–99%). The rate of the acetylation reaction is increased as the total acidity is increased, which is in the order $Fe^{3+} \approx Zn^{2+} > H^+ > K10 > Al^{3+}$; the activity is thus correlated to the acidity of the clay catalyst.

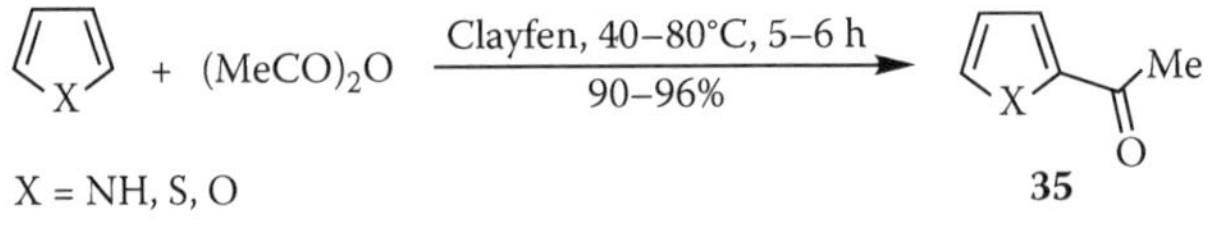

X = NH, S, O

35

Scheme 4.24

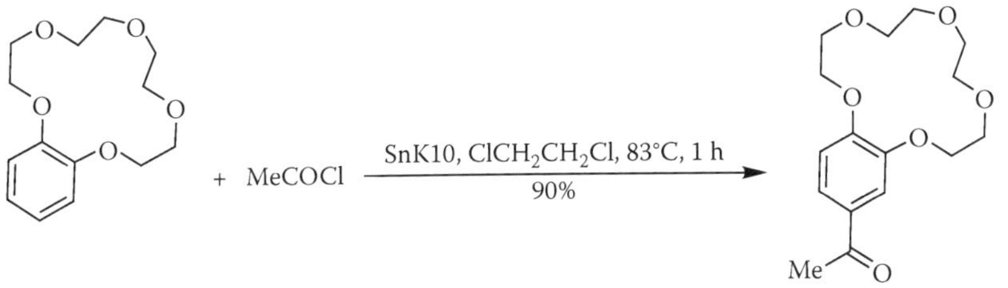

$$\text{36} + \text{MeCOCl} \xrightarrow[\text{90\%}]{\text{SnK10, ClCH}_2\text{CH}_2\text{Cl, 83°C, 1 h}} \text{37}$$

Scheme 4.25

$$\text{MeCOCl} + \text{Sn}^{2+} \longrightarrow \text{MeCOCl}^{+\cdot} + \text{Sn}^{+}$$

$$\text{MeCOCl}^{+\cdot} \longrightarrow \text{Me}\overset{+}{\text{C}}{=}\text{O} + \text{Cl}^{\cdot}$$

$$\text{Cl}^{\cdot} + \text{Sn}^{+} \longrightarrow \text{Cl}^{-} + \text{Sn}^{2+}$$

Scheme 4.26

Metal-exchanged K10 clays were studied as catalysts in the acylation of benzo crown ethers **36** with AC (Scheme 4.25).[139] The best catalyst is SnK10, which affords product **37** in 90% yield after 1 h. The activity of SnK10 is about three times higher than that of clayfen. Product **37** is accompanied by a minor by-product (~5%) due to the opening of the polyether ring followed by O-acetylation. The catalyst activity is lowered upon recycling. The redox mechanism depicted in Scheme 4.26 can represent an alternative pathway for the formation of an acyl cation that can account for the high activity of SnK10.

Gallium–magnesium hydrotalcite-like material (Ga-MgHT) with magnesium:gallium molar ratio of 3.0[140] shows a magnesium-oxide-like diffraction pattern, suggesting that gallium may be replacing some magnesium ions in magnesium oxide. The fresh Ga-MgHT shows no catalytic activity in the benzoylation of toluene with BC. However, after treatment with gaseous hydrogen chloride, the resulting material gives 90% conversion of BC after 3 h.[141] The hydrotalcite is activated by drying at 80°C, and the sample is designated as Ga-MgHT-80. Hydrogen chloride pretreatment of Ga-MgHT-80 is performed by bubbling gaseous hydrogen chloride through a mixture of the catalyst and the aromatic substrate, and then flushing with nitrogen to remove the physically adsorbed or absorbed hydrogen chloride from the reaction mixture. After this pretreatment, the surface area of the catalyst decreases from 9.0 to 2.9 m² × g⁻¹. The activity values measured in terms of the time required for half the reaction for different aromatic substrates follows the order mesitylene > *para*-xylene > toluene. The catalyst can be reused repeatedly in the reaction even in the presence of moisture, after washing with the hydrocarbon. The higher activity of the hydrogen-chloride-treated hydrotalcite is ascribed to the formation of highly dispersed gallium(III) and magnesium chlorides on magnesium oxide. These positive results, however, must be considered with due caution since no information on possible leaching of active species into solution has been reported.

The examples described here show that metal-exchanged clays can be easily prepared and efficiently utilized in the electrophilic acylation of different aromatic compounds carrying various functionalities. However, in our opinion, more studies are needed to give convincing evidence of the efficient reusability and the real heterogeneity of these catalysts.

4.3 Metal oxides

Metal oxides and their composite derivatives show catalytic activity and synthetic utility in fundamental organic reactions, including Friedel–Crafts acylation. Metal oxides are particularly useful as catalysts since they are readily available, inexpensive, and generally noncorrosive and nonhygroscopic. Moreover, a great number of methods are available for the preparation of single and mixed metal oxides that enable synthetic chemists to achieve the optimum catalytic performance by tailoring the physicochemical properties such as porosity, surface area, acidity, redox properties, and so on.[142] Finally, metal oxides can be utilized as robust supports for both organic and inorganic acid catalysts anchored through covalent as well as noncovalent bonds.

Benzoylation of the three isomeric xylenes is performed with BC in the presence of different metal oxides, namely, zinc oxide, iron(III) oxide, tin(II) oxide, and molybdenum(VI) oxide, at 150°C. The corresponding dimethylbenzophenones are produced in high yield (81%–97%), iron(III) oxide being the best promoter.[143]

Benzoylation of anisole with BC at the para-position is performed in the presence of a catalytic amount of iron(III) oxide (85% yield); lower yields of 4-halobenzophenones are achieved with the weakly deactivated chlorobenzene (33%), bromobenzene (31%), and iodobenzene (30%). However, taking into account the possible leaching of active iron trichloride produced by the reaction of iron(III) oxide with hydrogen chloride, the contribution of the homogeneous iron-based Lewis acid form is strongly suspected.

The acylation of toluene with BC over ferrous and ferric sulfates previously heated at 500°C, 700°C, and 900°C is described;[144] the catalysts prepared by calcination of the two sulfates at 700°C shows a good activity (ferrous sulfate: initial rate = 7.7×10^3 [mol $\times$ l^{-1} $\times$ min^{-1} $\times$ g^{-1}]; ferric sulfate: initial rate = 10.1×10^3 [mol $\times$ l^{-1} $\times$ min^{-1} $\times$ g^{-1}]). The product distribution is 18%–22% *ortho*-, 2%–4% *meta*-, and 74%–78% *para*-methylbenzophenone.

Table 4.14 Acylation of aromatics with acyl chlorides in the presence of zinc oxide

R^1	R^2	R^3	R^4	R^5	t (min)	Yield (%)
H	H	OMe	H	Ph	5	95
H	H	Me	H	Ph	10	86
CH=CH-CH=CH	H	OMe	Ph	10	67
H	H	H	H	Ph	120	50
H	H	OMe	H	Bn	5	98
H	OMe	H	OMe	Bn	5	92
H	H	Cl	H	Bn	5	85
H	H	OMe	H	Me	5	97
H	OMe	H	OMe	Me	10	90
H	H	Cl	H	Me	10	87
H	H	OMe	H	2-ClC$_6$H$_4$	10	95

Both ferrous and ferric sulfates are mainly remaining in a sulfate form when calcined at 500°C and decomposed, giving iron oxides containing 0.15% of sulfur at 700°C.

Zinc oxide, an inexpensive and commercially available inorganic solid, can be utilized as an efficient catalyst in the Friedel–Crafts acylation of activated and unactivated aromatic compounds with acyl chlorides at room temperature for 5 to 120 min (Table 4.14).[145] Acylation is claimed to occur exclusively at the para-position of the monosubstituted aromatic compounds. The catalyst can be recovered and reused, after washing with methylene chloride, for at least two further cycles, showing quite similar high yield (~90%) in the model benzoylation of anisole. Mechanistically, it seems that zinc chloride can be the true catalyst, generated in situ by the reaction of zinc oxide with hydrogen chloride.

Different metal oxides (silica, alumina, and titania) are utilized as solid catalysts in the high-temperature (~400°C) acylation of 2-methyl-imidazole **38** with AAN under a nitrogen stream (Scheme 4.27).[146] 4-Acetyl-2-methylimidazole **39** is produced in 65% yield in the presence of silica. The method is applied to the large-scale preparation of acylated imidazoles variously substituted on the heteroaromatic ring.

Among the various metal oxides studied as solid catalysts in Friedel–Crafts acylation reactions, cerium oxide gives results of some interest.[147] The catalyst is prepared by the precipitation method from cerium

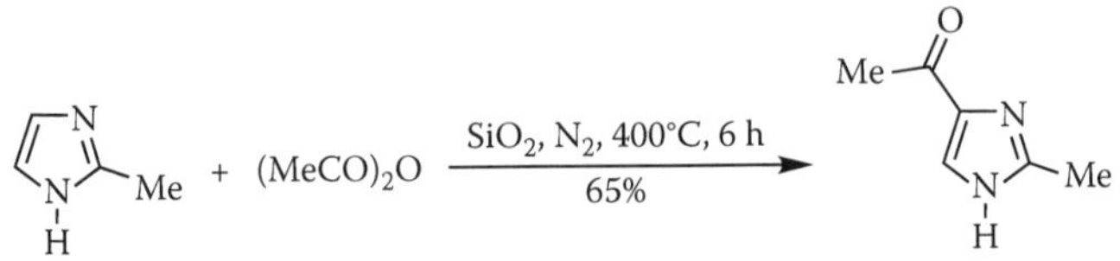

Scheme 4.27

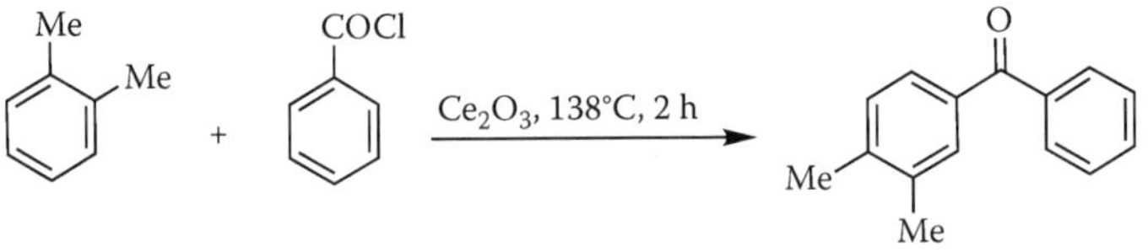

Scheme 4.28

nitrate and ammonia. The hydroxide thus produced is then calcined at a temperature between 100°C and 800°C for 2 h. The benzoylation of *ortho*-xylene with BC is studied with this catalyst at 138°C (Scheme 4.28). The main product is 3,4-dimethylbenzophenone accompanied by a small amount of 2,3-dimethylbenzophenone. The catalyst is effective for the benzoylation reaction upon activation by heating at 500°C. Even though this catalyst has a very low surface area (2.82 m$^2 \times$ g^{-1}) and pore volume (0.0030 cm$^3 \times$ g^{-1}), the presence of strong acid sites on its surface appears to be very important for the conversion of BC into the electrophile PhCO$^+$.

An efficient and ecofriendly procedure for the small-scale acylation of ferrocene with carboxylic acids is based on the in situ production of the mixed carboxylic-triflic anhydride (Table 4.15; see also Chapter 3).[148] The reaction is simply performed by adsorbing ferrocene on the surface of activated alumina (preheated at 150°C for 3 h) and adding a mixture of carboxylic acid and trifluoroacetic anhydride (TFAA) at room temperature for a selected time. Products **40** are recovered in 55%–98% yield simply by elution with diethyl ether.

The same methodology can be applied to the acylation of activated aromatic and heteroaromatic compounds.[149] A large variety of aromatic and aliphatic carboxylic acids are utilized as acylating reagents, giving the corresponding ketones in high yields (60%–96%). Although the use of TFAA for the acylation of aromatic substrates with carboxylic acids has been reported (see Chapter 3), this study shows a great generalization. In addition, the present procedure offers significant improvements to the previous methods such as reduction of the reaction time (10 min compared to several hours), minimization of the amount of TFAA (from 3–4 to

Table 4.15 Ferrocene acylation with carboxylic acid–TFAA
mixtures promoted by alumina

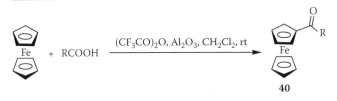

R	t (h)	Yield (%)
Me	0.5	98
Et	0.5	90
Bn	1.5	96
C_7H_{15}	1.0	93
$C_{17}H_{35}$	1.0	94
$MeCH(NO_2)(CH_2)_2$	1.0	85
$PhS(CH_2)_3$	1.5	88
Pr^i	0.5	95
$(Ph)_2CH$	1.5	93
$PhSCH_2CH(Me)$	1.5	89
$c\text{-}C_6H_{11}$	1.0	92
Ph	8.0	55
$4\text{-}OMeC_6H_4$	8.0	58

1.5 equivalents), and increase in the yields (the most marked examples are thiophene and furan, whose yields raise from 50% and 39% to 95% and 75%, respectively).

Gallium(III) oxide supported on MCM-41 mesoporous silica shows high catalytic activity with little or no moisture sensitivity in the acylation of aromatics with acyl chlorides.[150] The catalyst is utilized in 1,2-dichloroethane at 80°C for 3 h with different aromatic compounds, and aromatic as well as aliphatic acyl chlorides, giving ketones in 54%–82% yield. The activity order of the aromatic substrates is benzene (43% yield) < toluene (50% yield) < mesitylene (71% yield) < anisole (79% yield), in agreement with the electrophilic substitution trend previously observed.[15] This acylation reaction follows a probable redox mechanism similar to that described in Scheme 4.26.[139]

Thallium oxide supported on different materials was studied as a catalyst in benzene acylation with BC.[151] In general, TlO_x deposited on low-surface-area supports shows high catalytic activity (80% BC conversion after ~1.5 h), whereas that deposited on high-surface-area supports has almost no activity. This behavior is indicative of a different interaction between TlO_x and the supports. Strong metal-support interactions are well

Table 4.16 Acylation of aromatics with acetyl and benzoyl chlorides in the presence of zirconia

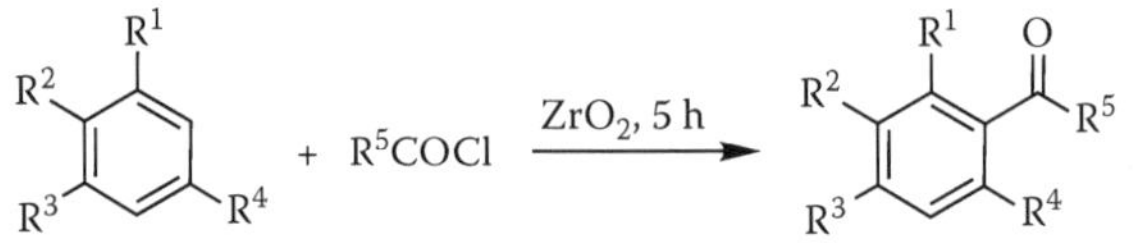

R¹	R²	R³	R⁴	R⁵	Solvent	T (°C)	Yield (%)
H	H	NHCOMe	H	Me	ClCH$_2$CH$_2$Cl	60	80
H	H	NHCOMe	H	Ph	—	120	71
H	H	OMe	H	Ph	—	120	65
CH=CH-CH=CH	H		OMe	Me	ClCH$_2$CH$_2$Cl	60	60
CH=CH-CH=CH	H		OMe	Ph	—	120	62
CH=CH-CH=CH	H		H	Ph	—	120	80
Me	H	OMe	OMe	Ph	—	120	80

known and are observed for many supported metals;[142,152] however, the fine analysis of the chemical nature of the strong metal-oxide-support inter-action is scarce.[153] In the present case, since TlO$_x$ is basic, it can chemically interact with the weak or strong acid hydroxy groups of high-surface-area supports. The low-surface-area supports are highly sintered macroporous materials, and hence, have only few surface hydroxy groups; thus, the supported TlO$_x$ keeps its catalytic properties unchanged.

Acylation of electron-rich arenes with AC, BC, and benzotrichloride can be performed in the presence of hydrated zirconia.[154] The catalyst is prepared by treatment of an aqueous solution of zirconium(IV) oxychlo-ride octahydrate (ZrOCl$_2$ × 8H$_2$O) with aqueous sodium hydroxide at room temperature, followed by heating the precipitate at 300°C for 5 h. The acety-lation is performed in 1,2-dichloroethane at 60°C, whereas benzoylation is carried out under solventless conditions at 120°C (Table 4.16). It is remark-able that naphthalene is benzoylated selectively at the 1-position. The catalyst is recovered by filtration and reused three times with no loss of activity and selectivity in all cases.

Thermally stable tungstated zirconias (WZ), prepared by adding aqueous ammonia solution to a mixed aqueous solution of zirconium(IV) oxychloride octahydrate and ammonium metatungstate, and calcining the aged coprecipitate at 700°C for 3 h, were tested in the model acetyla-tion of anisole with AAN.[155] Both anisole conversion and yield of 2- and 4-methoxyacetophenone at 50°C increase with the amount of tungsten(IV) oxide, the maximum activity (38% yield) being observed for 19% weight. On the other hand, the impregnated sample with 20 wt% tungsten(IV) oxide is catalytically inactive. The catalytic activity of the coprecipitated sample decreases with a higher amount of tungsten(IV) oxide (25 wt%:

Table 4.17 Activity of different catalysts in the benzoylation of
anisole with BC in nitrobenzene at 150°C for 22 h

Catalyst	Anisole conversion (%)	Yield (%)	Product distribution (%)		
			4-MBP	2-MBP	PB
BZ	91	88	94	3	3
SZ	96	94	95	3	2
BEA	95	93	95	3	2
AlCl$_3$	83	81	93	4	3

catalytic yield = 31%). This can be due to the surface acid site blockage by the separate tungsten(IV) oxide phase. The catalytic activity of the 19 wt% WZ improved at higher reaction temperatures. At 80°C, 74% of AAN is converted with 63% yield of methoxyacetophenones, and 82% yield is observed at 100°C.

The Brönsted acid site FT-IR band intensity (at 1491 cm^{-1}) directly correlates with the catalytic activity, and the maximum activity observed for 19 wt% WZ is due to the maximum number of Brönsted acid sites generated by the higher interaction of tungsten(IV) oxide with zirconia. Similar composite materials were studied in the acylation of 2-MN with AAN to selectively give the 1-acetylated product in 50% yield.[156]

Borate zirconia (BZ) with 30% mol of boron oxide is reported to be a strong solid acid catalyst (H$_0$ = −13.6).[157] This material is prepared by treatment of zirconium(IV) hydroxide with an aqueous solution of boric acid, followed by calcination of the resulting product at 550°C overnight.[158] The catalyst was studied in the benzoylation of anisole with BC at 150°C in nitrobenzene and compared with other heterogeneous [sulfate zirconia (SZ), BEA zeolite] and homogeneous (aluminum chloride) catalysts (Table 4.17). After 22 h, the conversion of anisole and the selectivity to 4-methoxybenzophenone in the case of BZ and other solid catalysts are higher than those achieved with an aluminum chloride catalyst; phenyl benzoate is the major side product. It must, however, be underlined that the catalytic activity of BZ is significantly lower when the benzoylation reaction is performed at a low temperature. SZ are superior for benzoylation and give high conversion at temperatures as low as 40°C.[159] This difference can be rationalized on the basis of surface acidity data; in fact, there is no difference between SZ and BZ in terms of a surface acid site concentration (0.35–0.40 mmol × g^{-1}) significantly higher than that found for pure zirconia, which does not show any catalytic activity in this reaction. However, SZ contains significantly stronger acid surface sites, which are responsible for the higher catalytic activity at 40°C.

An iron-silicate-type catalyst (SiFe) can be prepared by the reaction of sodium trisilicate and ferric nitrate in the presence of ammonia and

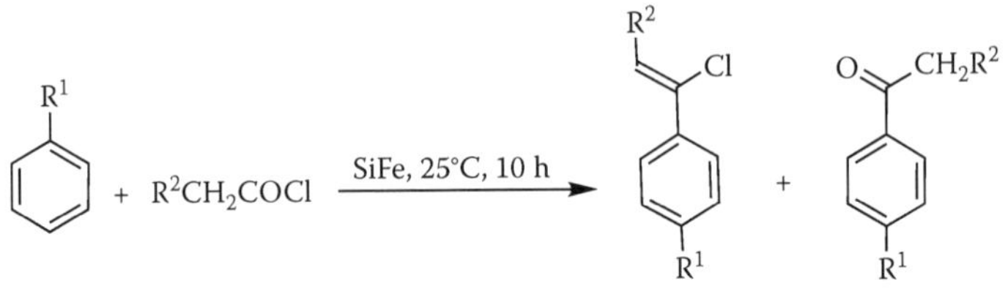

Scheme 4.29

ammonium carbonate, followed by calcination of the precipitate at 500°C. Friedel–Crafts acylation of anisole with BC in the presence of this catalyst at 25°C for 10 h without solvent gives *para*-methoxybenzophenone in 70% yield.[160] Different aromatic acyl chlorides react under the same conditions with anisole and thiophene, affording the corresponding aryl- and heteroaryl ketones in 70%–79% yield.

The same catalyst promotes the reaction of toluene and anisole with propanoyl chloride; the acylation product **42** (R^1 = OMe R^2 = Me) and 1-chloro-1-(4-methoxyphenyl)propene **41** (R^1 = OMe R^2 = Me) are obtained in 28 and 65% yield, respectively (Scheme 4.29).

The more complex polymetal oxides such as $Cu_xMn_{(1-x)}Fe_2O_4$-type ferrospinel systems (x = 0.0, 0.2, 0.4, 0.6, 0.8, and 1.0), are utilized as solid catalysts in the benzoylation of benzene with BC.[161] The catalysts are prepared by a low-temperature coprecipitation of a water solution containing the calculated amounts of the corresponding nitrate salts at pH 9–10, using 5 M sodium hydroxide solution; the precipitate is then calcined at 500°C. Comparison of the rate constant of the benzoylation with surface acidity and catalyst composition reveals that replacement of manganese(II) with copper(II) first decreases the rate constant and then shows an increase. The results are indicative of a relationship between the concentration of strong acid sites on the catalyst surface with the rate of benzoylation. The relatively high catalytic activity of copper(II)-substituted systems is due to the generation of Lewis acid sites at higher copper(II) loadings.

4.4   Sulfated zirconia

Sulfated metal oxide catalysts represent a class of extremely attractive strong solid acids showing widespread application in different areas of chemical transformations.[162] It was reported that sulfated zirconia (SZ) prepared by treatment of zirconia with sulfuric acid or ammonium sulfate exhibits extremely strong acidity, and it is able to catalyze the isomerization of butane to isobutane at room temperature.[163,164] This behavior

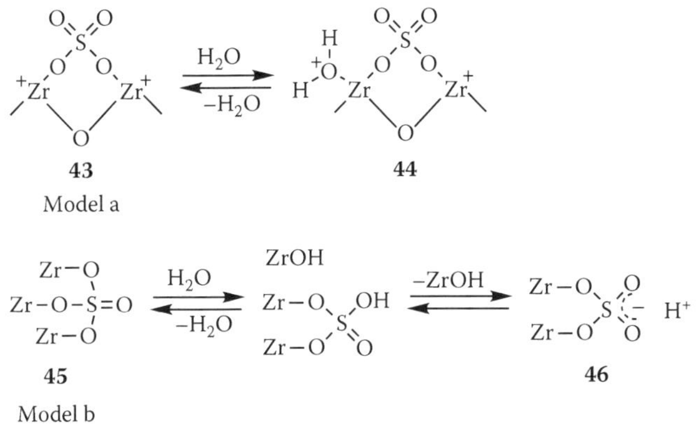

Scheme 4.30

represents a unique catalytic performance compared to typical solid acids such as zeolites, which show no activity for this reaction at such low temperatures. Using Hammet indicators, it was claimed that SZ is an acid 100 times stronger than 100% sulfuric acid;[164] therefore, SZ with $-H_0 = 16$ is considered the strongest halide-free solid superacid.[165] However, the superacidity of SZ has been recently questioned,[166] and SZ is claimed to be only a strong solid acid with an acid strength comparable to that of sulfuric acid or some acid zeolites.

Among the different sulfated metal oxides available, SZ is, by far, the most studied and utilized solid acid catalyst; its properties strongly depend on the preparation method (from the nature of the starting materials to calcination conditions).

A great number of studies were aimed at achieving information on the nature and the origin of the intense acidity of SZ. It was proposed that strong acidity is generated by the electron inductive effect of the S=O double bond, which increases the electron-deficient nature of the metal atom attached to the sulfate group, thus strengthening its Lewis acid character (structure **43** in Scheme 4.30 Model a). However, since SZ catalyzes many reactions in which a proton addition represents the initial and crucial step, structure **44** may represent the Brönsted acid site. The Lewis acid sites predominate in the absence of water and after calcination at high temperature; in the presence of water, Lewis acid sites are converted into Brönsted acid sites via proton transfer. Some other models were suggested to account for the acid properties of SZ, all based on the possible transformation of the Lewis into Brönsted acid sites (equilibrium **45** ⇄ **46** of Scheme 4.30 Model b).

In an early study,[167] it was shown that SZ calcined at 650°C was catalytically active in the acylation of toluene with acetic and benzoic acids at 180°C. In the latter case, the yield of methylbenzophenones is ~60%, and the product distribution is 30%–35% *ortho-*, 5% *meta-*, and 60%–65% *para*-methylbenzophenone. The same reaction carried out with BAN takes place readily even at 30°C, and it seems also to be accomplished by the BAC produced. The real heterogeneity of the catalyst is confirmed by the inertia observed when the catalyst is removed from the reaction medium. The acylation of toluene with AAC is also performed in a flow system by passing the vaporized reaction mixture through the SZ catalyst bed at 280°C with nitrogen carrier. The system seems to be quite promising even if no further development of the research has yet been performed (51% AAC conversion; 89% selectivity with respect to all ketones; ortho-, meta-, and para-isomer selectivity 16%, 13%, and 71%, respectively).

A commercially available type of SZ, calcined at 550°C, was found to be an efficient catalyst in the bezoylation of toluene with BAN at 100°C for 3 h.[168] The methylbenzophenones yield is 92% (ortho/para = 0.37).

As already observed in different reactions, the SZ preparation procedure controls the catalytic activity. Indeed, the catalyst prepared by treatment of zirconium(IV) hydroxide with sulfuric acid followed by calcination at 600°C gives the methylbenzophenones in 25% yield, whereas 52% yield is achieved with SZ prepared by hydrolysis of zirconium(IV) oxychloride with sulfuric acid, followed by calcination at 550°C.

This catalyst was also studied in the benzoylation of 1-methoxynaphthalene (1-MN) with BAN at 40°C to afford the 4-benzoyl-1-MN.[169] Comparative results with classic solid acids (Table 4.18) confirm the higher activity of SZ with respect to solid acids containing a greater number of acid sites under the same conditions.

In a detailed comparative kinetic study carried out with toluene and BAN, it is shown that the activity of pure Nafion is about half of that of SZ. However, the isomer distribution is different: the ortho/para acylation ratio is 0.25 for Nafion, whereas it is 0.34 for SZ.[170]

SZ can be efficiently applied to the acylation of anisole, with a variety of anhydrides (Table 4.19), affording the para-isomers **47** in high yield, accompanied by a small amount of the ortho-isomers **48**.[171,172]

In addition, electron-rich as well as electron-poor aromatic compounds react with BAN in the presence of and, generally, high yields of the desired aromatic ketones are obtained (Table 4.20).[159,172] By using chiral anhydrides such as (S)-2-methylbutyric anhydride, the pure (S)-1-(4-methoxyphenyl)-2-methylbutan-1-one is isolated in 95% yield (Scheme 4.31).[171]

Similar good synthetic results are achieved by using acyl chlorides as acylating agents (Table 4.21). In this case, too, the para-isomers **49** are obtained in higher yield and selectivity, the ortho-isomers **50** being obtained in lower amounts.

Table 4.18 Activity of different heterogeneous catalysts in the benzoylation of 1-MN with BAN in ClCH$_2$CH$_2$Cl at 40°C for 4–20 h

Catalyst	Surface area (m^2/g)	Pore diameter (Å)	Acid sites (mmol/g)	Yield (%)
SZ	150	68	0.41	50
Cs$_{2.5}$H$_{0.5}$[PW$_{12}$O$_{40}$]	60	72	0.16	40
Nafion/SiO$_2$	224	220	0.10	22
Amberlyst-15	38	400	4.7	19
K10	233	60	0.28	13
BEA	500	7.6 × 6.4	0.87	4
MOR	400	7.0 × 6.5	0.76	3

Table 4.19 Anisole acylation with anhydrides in the presence of SZ

		Product distribution (%)		
R	Yield (%)	**47**	**48**	Reference
Me	91	98	2	171
Et	92	98	2	171
Pri	96	99	1	171
Bus	95	99	1	171
C$_9$H$_{19}$	90	98	2	171
C$_{17}$H$_{35}$	88	98	2	171
Ph	95	96	4	172

Polycyclic aromatic compounds such as naphthalene, methoxynaphthalenes, dimethylnaphthalenes, and anthracene undergo efficient acylation with acyl chlorides and anhydrides in the presence of the SZ produced from zirconium(IV) oxychloride by a methodology similar to that previously mentioned.[169]

Table 4.20 Aromatic benzoylation with BAN in the presence of SZ

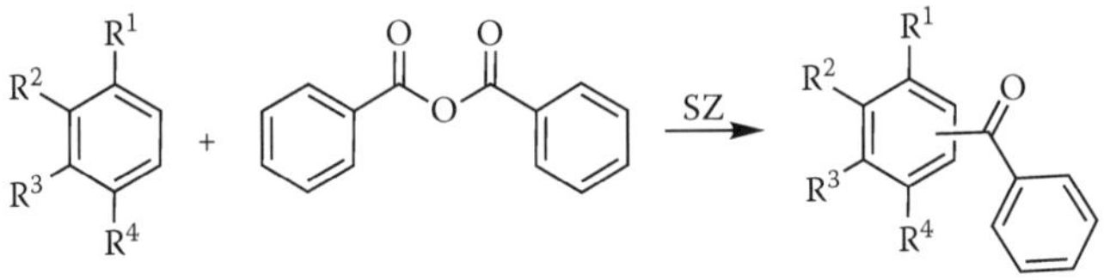

R^1	R^2	R^3	R^4	Yield (%)	Product distribution (%)	t (h)	T (°C)	SZ^a (g)	Reference
Me	H	Me	H	92	2,4-Me2-BP (91)	18	100	75	159
					2,6-Me2-BP (9)				
Me	H	H	Me	31	2,5-Me2-BP (100)	24	110	75	159
OMe	H	Cl	H	85	2-Cl,4-OMe-BP (84)	20	120	112	172
					4-Cl,2-OMe-BP (9)				
					2-Cl,6-OMe-BP (7)				
OMe	Cl	H	H	70	3-Cl,4-OMe-BP (100)	5	120	112	172

^{a} Grams of SZ per millimole of (PhCO)$_2$O (limiting reagent).

1-MN can be acylated with AAN and benzoylated with BAN in nearly 100% yield, giving 1-acyl-4-MN. BC reacts more slowly (82% yield).

The acylation of 2-MN occurs at 1-position and does not exceed 85%; no enhancements are obtained either by increasing the amount of SZ or the reaction temperature or by extending the reaction time.

1,3-Dimethylnaphthalene reacts easily and selectively over SZ with BAN and BC to give 1-benzoyl-2,4-dimethylnaphthalene in 93% and 79% yield, respectively. Moreover, anthracene is easily and selectively converted into 9-benzoylanthracene in 61% yield with BAN, and 67% yield with BC (Scheme 4.32).

Results from Tables 4.19–4.21 allow us to draw the important conclusion that electrophilic aromatic acylation can be performed with SZ catalysts containing a substoichiometric number of catalytically active surface sites in comparison to the alternative classic Friedel–Crafts acylation reactions. However, further and deeper studies are desirable on this

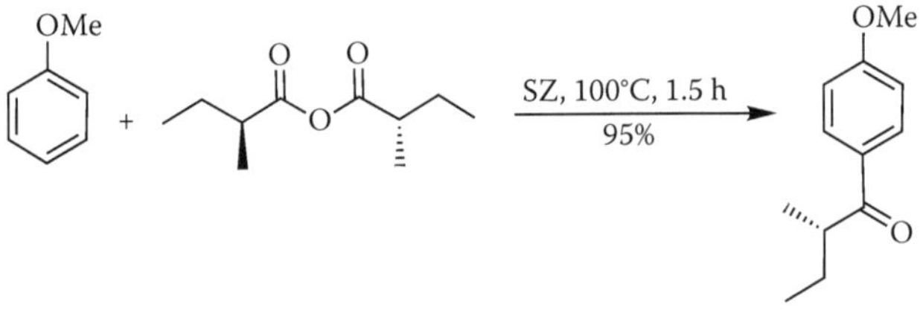

Scheme 4.31

Table 4.21 Anisole acylation with acyl chlorides in the presence of SZ

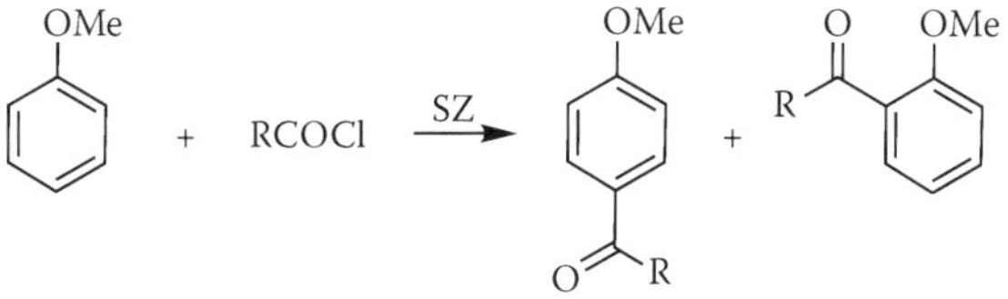

					Product distribution (%)	
R	SZ[a] (g)	t (h)	T (°C)	Yield (%)	49	50
Ph	30	15	100	92	96	4
Bz	30	15	100	90	96	4
$Ph(CH_2)_2$	30	15	100	92	97	3
$MeOCO(CH_2)_2$	75	1.5	100	31	96	4
$4\text{-}MeC_6H_4$	30	15	110	92	96	4
$4\text{-}ClC_6H_4$	30	15	110	91	92	8
$4\text{-}MeOC_6H_4$	30	15	120	87	97	3
$4\text{-}NO_2C_6H_4$	30	15	140	60	82	18

[a] Grams of SZ per millimole of RCOCl (limiting reagent).

very promising topic with particular emphasis on catalyst lifetime and possible catalyst recycling.

Some occasional information is available concerning the reusability of the SZ catalyst in the benzoylation of anisole with BAN. The catalyst is effectively utilized for a second cycle, but more drastic experimental conditions are needed to achieve a yield comparable to that of the first run. Moreover, detailed spectroscopic analysis of the utilized catalyst confirms that, during the reaction, carbonaceous deposits are formed on the surface, and the number of the available acid surface sites decreases.

A similar trend is observed in the benzoylation of benzene with *para*-chlorobenzoyl chloride over SZ catalyst prepared from zirconium(IV) oxychloride and calcined at 650°C for 3 h (Scheme 4.33).[173]

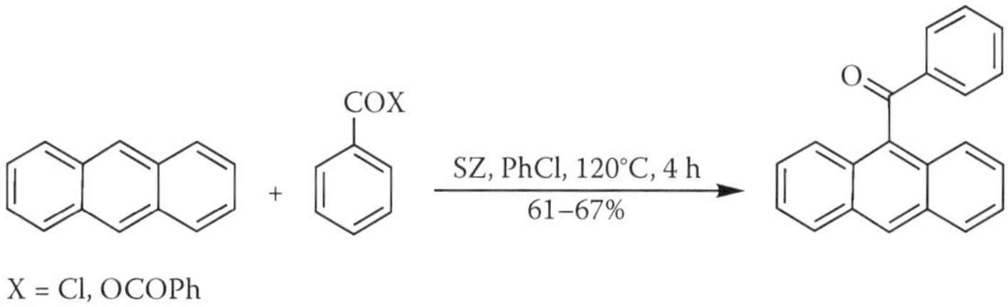

X = Cl, OCOPh

Scheme 4.32

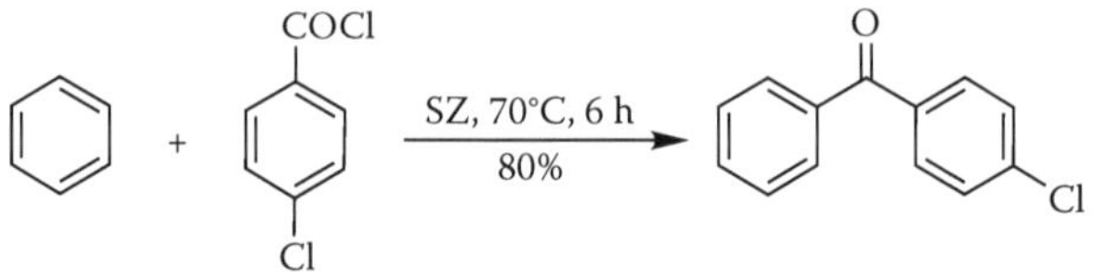

Scheme 4.33

para-Chlorobenzophenone is the sole reaction product obtained with 80% yield. The catalyst can be recycled after filtration. However, since there is inevitably loss of particles during the filtration process, the subsequent cycle gives lower yield; however, accounting for the loss of about 20% of catalyst in the recovering procedure, the initial rate drop is only about 10% in the subsequent run.

Positive results on catalyst SZ reusability are reported and show that the catalytic activity of spent SZ can be completely restored by burning the deposited carbonaceous material.[174] In the benzoylation of anisole with BAN, the conversion of BAN is almost complete after 30 min. The catalyst is reused after washing, exhibiting a slightly lower initial activity. However, catalytic activity can be completely restored by calcining the used catalyst at 550°C for 15 min. Detailed spectroscopic studies confirm the presence of benzoate species on the catalyst surface. The acylating agent and the benzoic acid formed during the reaction may be regarded as the origin of benzoate species, which may be formed under the involvement of nonacidic surface hydroxyl groups. The population of nonacidic surface hydroxyl groups on the surface of SZ depends on the preparation method and can be minimized by applying preparation routes that ensure a more homogeneous deposition of surface sulfate species such as aerogel synthesis. Actually, in a comparative study with a series of zirconias, it was confirmed that aerogel SZ samples reveal higher concentration of acid sites and larger surface areas, namely, a mean concentration of acid sites between 2 and 3 μmol × m^{-2}.[175] These results clearly indicate the dominating impact of Brönsted acid sites for the benzoylation of anisole over modified zirconia catalysts. It is suggested that, under these conditions, the benzoylation reaction only occurs on the Brönsted acid sites (model **51** in Figure 4.9). In contrast, the Lewis acid sites are not able to produce the active acylating species (model **52** in Figure 4.9) and, consequently, little or no conversion of the benzoylation reagent into the desired aromatic ketone is observed.

A second possible mechanism of aromatic acylation on catalysts containing sulfate or sulfonate groups such as SZ is depicted in Scheme 4.34. This mechanism is related to the tendency of these catalysts to form surface-bounded sulfate or sulfonate anions as thermodynamically favored leaving groups. Thus, formation of surface-bounded

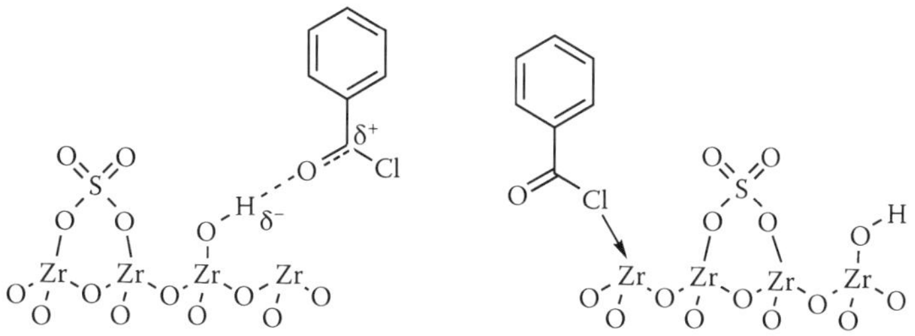

51   52

Figure 4.9 Possible interactions between BC and SZ.

acylsulfates or acylsulfonates as acylating agents in the catalytic cycle can be assumed.[169]

With the aim of producing more active, stable, and regenerable catalysts, iron- and manganese-promoted SZ (FMSZ) were prepared by incipient wetness impregnation, starting from zirconia and ammonium sulfate to give the sulfated zirconia, which is successively stepwise impregnated with ferric nitrate and manganese(III) nitrate solutions.[176] The catalysts are calcined at 500°C before use.[177] The mechanistic cycle of the Friedel–Crafts acylation reaction is proposed to incorporate both metal ions as redox sites (to attract the reactants) and Brönsted acid sites (to activate the acylating reagent). In the benzolyaltion of anisole with BC, the yield of 4-methoxybenzophenone considerably increases from 25% to 95% when the temperature is raised from 40°C to 100°C.

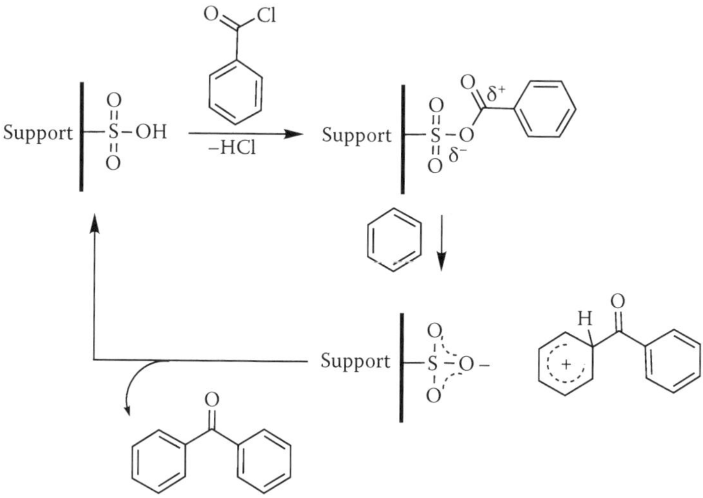

Scheme 4.34

In the benzoylation of toluene with BC, an increase of iron content results in the increase of conversion and catalytic activity.[177,178] The high activity of the iron-promoted system may be attributed to the redox properties of iron as well as its strong acidity. In particular, in this case, Lewis acidity seems to play a crucial role in the catalytic activity. In fact, Lewis-acidity enhancement, determined by perylene adsorption studies, parallels the steady increase in reactivity observed with the increase in the percentage of iron content.

Benzoylation of benzene, toluene, and *ortho*-xylene at the corresponding refluxing temperatures for 1 h over iron SZ (2% iron) follows the expected order: benzene (15%) < toluene (80%) < *ortho*-xylene (96%). In the model reaction between BC and toluene, the truly heterogeneous nature of the catalyst as well as its resistance to deactivation during reuse is confirmed.

Detailed spectroscopic and microcalorimetric studies suggest that, in the aluminum-promoted SZ, the surface enrichment in aluminum and the formation of aluminum–oxygen–zirconium bonds on the surface are probably the cause for enhancement in the amount of intermediate strong acid sites.[179,180] According to the principle of electronegativity equalization,[181] since the electronegativity of aluminum is higher than that of zirconium(IV), the positive charge on zirconium atoms is increased because of the formation of aluminum–oxygen–zirconium bonds. At the same time, the charge from zirconium atoms to the neighboring aluminum atoms strengthens the aluminum–oxygen bond between the surface sulfate species. The stronger aluminum–oxygen bond leads to an increase in the thermal stability of the surface sulfate species and, consequently, the acidity of the catalyst is enhanced. This catalyst, containing 4% alumina, gives 91% yield of methylbenzophenones in the benzoylation of toluene with BC at 110°C for 10 h.

The activity of sulfated alumina (SA), prepared by impregnating γ-alumina with sulfuric acid, and SZ, both calcined at 650°C, is compared in the same reaction.[182] The activity of SA, measured through BC conversion at 110°C after 1 h, is lower (45%) than that of SZ (78%).

A double metal oxide sulfate solid superacid (alumina-zirconia/persulfate, SA-SZ) can be prepared by treatment of a mixture of aluminum hydroxide and zirconium(IV) hydroxide with an aqueous solution of ammonium persulfate, followed by calcination at 650°C.[183] This catalyst can be efficiently utilized in the benzoylation of arenes with benzoyl and *para*-nitrobenzoyl chloride (Table 4.22), giving BPs in interesting yields. Even if 1 g of catalyst is needed for 40 mmol of chloride, the process seems to be quite useful because the catalyst can be readily regenerated by heating after washing with acetone and diethyl ether and reused four times.

Table 4.22 Benzoylation of aromatics with BCs in the presence of SZ-SA

R¹	R²	R³	R⁴	R⁵	R⁶	T (°C)	t (h)	Yield (%)	Selectivity (%)
H	H	H	H	H	H	100	6	75	100
H	H	H	Me	H	H	110	5	62	80
H	H	Me	Me	H	H	100	4	85	100
H	Me	H	Me	H	H	100	4	87	100
H	Me	H	H	Me	H	100	4	80	96
H	Me	H	Me	H	Me	110	4	88	100
H	H	H	OMe	H	H	100	2	66	72
H	H	H	Cl	H	H	135	5	74	88
H	H	H	Br	H	H	150	5	77	91
H	H	Cl	Cl	H	H	140	5	81	100
NO₂	H	H	H	H	H	100	4	77	100
NO₂	H	H	Me	H	H	110	4	61	78
NO₂	H	Me	Me	H	H	100	3	87	100
NO₂	Me	H	Me	H	H	100	3	89	100
NO₂	Me	H	H	Me	H	100	3	85	100
NO₂	H	H	OMe	H	H	100	2	60	65
NO₂	H	H	Cl	H	H	135	4	72	85
NO₂	H	H	Br	H	H	150	4	76	87

Gallium(III)- and iron(III)-promoted SZ can be supported on mesoporous materials such as MCM-41 silica.[184] The catalysts, named GaSZ/MCM-41 and FeSZ/MCM-41, respectively, are prepared by incipient wetness impregnation using zirconium(IV) sulfate as the precursor in combination with gallium(III) nitrate or ferric nitrate and calcined at 700°C for 3 h. These catalysts are studied in the acylation of veratrole with AAN (Scheme 4.35). The reaction gives only 3,4-dimethoxyacetophenone, AAC being the sole detectable by-product. The most active catalyst is GaSZ/MCM-41 (78% yield, 100% selectivity), followed by SZ/MCM-41 alone (68% yield, 100% selectivity) in the reaction carried out at 80°C for 3 h, whereas the FeSZ/MCM-41 shows much lower activity (49% yield, 100% selectivity). IR studies confirm that not only Brönsted sites but also Lewis acid sites are effective in the activation of the acylating agent. In fact,

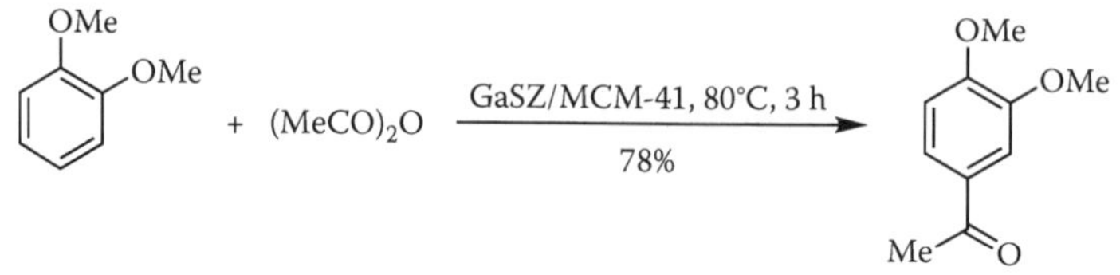

Scheme 4.35

the FT-IR C=O adsorption analysis shows that Lewis acidity is present in the case of GaSZ/MCM-41 and SZ/MCM-41 catalysts, whereas it is almost absent in the FeSZ/MCM-41 sample. The best catalyst GaSZ/MCM-41 can be recycled three times after washing and calcination in air at 450°C for 90 min, giving 78%, 76%, and 70% yield in the three cycles.

From the preceding experimental data and comments, it is clear that both Lewis and Brönsted acid sites exist on sulfated oxide catalysts, but the detailed role of the two types of acidity in various reactions, including Friedel–Crafts acylation, is still controversial.

4.5   Heteropoly acids (HPAs)

HPAs are Brönsted acids composed of heteropoly anions and protons as countercations; the most commonly utilized HPA is the phosphotungstic acid $H_3PW_{12}O_{40}$ (PW). HPAs are stronger than many conventional solid acids such as mixed oxides and zeolites. One important advantage of these catalysts is that they can be utilized both homogeneously and heterogeneously. The homogeneous reactions occur in polar media at ~100°C; on the other hand, when using nonpolar solvents, the reactions proceed heterogeneously. The HPA catalysts are easily separated from the homogeneous systems by extraction with water, and from the heterogeneous reaction mixture by filtration.

The use of HPAs and multicomponent polyoxometalates as catalysts in liquid-phase reactions was reviewed by Kozhevnikov.[185] Moreover, an interesting minireview was published concerning the Friedel–Crafts acylation of arenes and the Fries rearrangement catalyzed by HPA-based solid acids.[186] The results show that HPA-based solid acids, including bulk and supported heteropoly acids as well as heteropoly acid salts, are efficient and environmentally friendly catalysts for all reactions analyzed.

BBA cycloacylation has been studied in the presence of different polyoxometalate catalysts with strong acid character, using both a batch and a pseudocontinuous reactor.[187] Batch runs have been performed for screening the catalysts, whereas the continuous reactor was used with the best catalyst to obtain detailed information regarding kinetics, catalyst

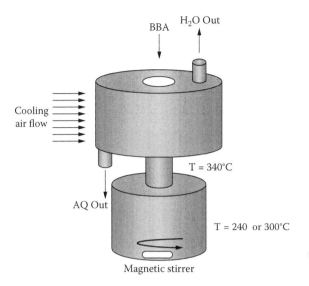

Figure 4.10 Scheme of the reactor/evaporator employed in the continuous-flow AQ preparation from BBA in the presence of hydrate $H_7PW_{12}O_{42}$ catalyst. (From Tesser, R., Di Serio, M., Ambrosio, M., and Santacesaria, *Chem. Eng. J.*, 90, 195, 2002. With permission.)

productivity, reaction yields, and catalyst deactivation. It was found that a synthesized polyoxometalate catalyst, hydrate $H_7PW_{12}O_{42}$, gives much better performances than those achieved with some previously tested catalysts such as acid bentonites and Y and BEA zeolites. Catalytic screening has been made at 200°C for 2 h. All the tested polyoxometalate catalysts show satisfactory yields in AQ (up to 29%). With the best catalyst, hydrate $H_7PW_{12}O_{42}$, continuous runs have been performed in the reactor/evaporator schematized in Figure 4.10.

Three zones in this reactor are working at different temperatures: (1) the reacting zone in which the reagent and catalyst are initially loaded and are kept at the reaction temperature (240°C or 300°C) with a furnace, (2) the neck, which is heated at temperatures of about 340°C to favor the diffusion of the vapors of AQ and water, and (3) the product-recovering zone that is cooled with a cold-air stream in which the sublimated AQ crystallizes. The activity and productivity of the polyoxometalates are higher than the ones observed on other solid acid catalysts, such as zeolites, and also in pseudocontinuous longer runs. In particular, the run performed at 300°C shows a productivity that is more than one order of magnitude higher.

It is well known that one of the major problems associated with HPAs in the bulk form is their low efficiency due to low surface area

(1–5 m^2 × g^{-1}),[185] possible deactivation, and relatively poor stability. Attempts to improve the efficiency and stability of HPAs were made by using various supports, including mesoporous silica,[188] mesoporous aluminosilicates,[189] zirconia, and alumina.[190]

Zirconia-supported PW (PW/ZrO$_2$) catalyst shows good activity in the acylation of diphenyl ether with BC.[191] The maximum conversion of BC is ~60%, and 4-benzoyldiphenyl ether is obtained with a selectivity higher than 97%. The catalyst can be efficiently recycled three times, affording the same BC conversion.

The acylation of anisole with AAN was examined in the presence of silica-supported PW.[192–194] The supported catalysts are prepared by impregnating silica (surface area: 300 m^2 × g^{-1}) or mesoporous silica MCM-41 (surface area: 1250 m^2 × g^{-1}) with a methanol solution of PW. The acylations are carried out in liquid phase in a glass reactor charged with aromatic substrate and AAN, the substrate taken in excess over the acylating agent; no solvent is used.

The control of water content in heteropoly acid catalysts proved essential for their efficient performance; this can be achieved by thermal pretreatment, which is typically done at 130°C–200°C.[185] In the present process, the optimum pretreatment temperature is 150°C at 0.1 torr. Apparently, these water molecules are hydrogen bounded to the acidic protons; the amount of AAN that may be consumed reacting with this water is negligible (ca. 1%). It must be underlined that an excess of water may cause a decrease in the HPA acid strength and thus in its catalytic activity; on the contrary, a strong dehydration of the catalyst increases the acid strength but decreases the number of acid sites, which will reduce the catalytic activity.

In the reaction of anisole with AAN at 110°C, the para-acylation process dominates, with only a small percentage of the ortho-acylation product when 40% PW/SiO$_2$ is utilized; the selectivity toward monoacylation is practically 100%. The acylation appears to be a truly heterogeneously catalyzed reaction: no contribution of homogeneous catalysis by PW is observed when the catalyst is filtered off at the reaction temperature. The catalyst is found to be reusable, although gradual decline of activity is observed; better results are obtained when, after the first run, the catalyst is filtered off and washed with methylene chloride. Apparently, this treatment removes tars more efficiently from the catalyst surface. Such a procedure allows 82% of the initial *para*-methoxyacetophenone yield to be obtained in the second run. Coking may cause partial deactivation of the catalyst, which is evident from the dark brown color of the catalyst, initially a white powder.

A comparison of the activity of two PW catalysts, one supported on a commercial silica (PW/SiO$_2$) and the other on a silica–zirconia-mixed oxide (PW/SiZr), was performed in the same reaction.[195] Both supported catalysts, and the SiZr support are active, even if the maximum yield (65%) is observed with PW/SiO$_2$, whereas low yields (~20%) are obtained with SiZr and PW/SiZr. Differences in the nature and number of the acid sites are recognized, and dispersion of PW over the silica surface results in the generation of a catalyst possessing Brönsted acid properties. However, addition of PW to a support with inherent Lewis and Brönsted acid properties modifies the number and distribution of the acid sites, usually by increasing their number and strength; these effects depend on the extent of interaction of the PW with the support. The interaction of PW with the silica surface, which is much weaker than that for zirconia,[196] is attributed to the formation of ($\equiv$SiOH$_2^+$)(H$_2$PW$_{12}$O$_{40}^-$) species and can account for the different activity of the two PW-supported catalysts.

The acylation of aromatics with crotonic acid can be carried out with pure and silica gel-supported HPA (Scheme 4.36).[197,198] The protonation of crotonic acid in the presence of a Brönsted acid produces an electrophilic reactive intermediate that can be either an alkyl or an acyl carbocation that can react with the aromatic ring to produce either the alkylated **53** or the acylated products **54,** respectively. Products **53** and **54** can further react either via intramolecular acylation/alkylation reaction to produce the corresponding indanones **55** and **57** or via an intermolecular acylation/alkylation process to form the corresponding ketones **56.** Products **55–57** can be obtained in different proportions, depending on the relative rates of the different reactions.

Analysis of products obtained with *para*- and *meta*-xylene shows that all the catalysts are more active for the acylation than the alkylation reaction. Indeed, indanones **58** and **59** (Figure 4.11) are obtained in 65% and 79% yield.

As already mentioned, a peculiarity in the use of these interesting catalysts is represented by their solubility in polar organic compounds such as aromatic ethers even in the case of silica-supported HPAs. These results suggest that further studies are needed to point out an efficient HPA-based solid catalyst for Friedel–Crafts acylation. Trying to overcome this problem, the alkali metal and ammonium salts of PW were studied in the acylation of *para*-xylene with BC.[199] The catalytic activity of sodium and potassium salts decreases with increasing the metal content, expressed as x in the formula M$_x$H$_{3-x}$[PW$_{12}$O$_{40}$]. For rubidium and cesium salts, however, the activity first decreases with increasing x, then jumps to attain a maximum at x = 2.5. Although the acid alkali metal salts, as well as the parent-free acid PW, are protonic acids in nature, they work as efficient catalysts to activate BC. In contrast, typical strong protonic acids such as sulfuric and perchloric acids are quite inactive for the present benzoylation reaction. The reason is that the heteropoly anion PW$_{12}$O$_{40}^{3-}$ has the ability to stabilize cationic intermediates[200] such as benzoyl cations, whereas simple oxoanions such as sulfate and perchlorate

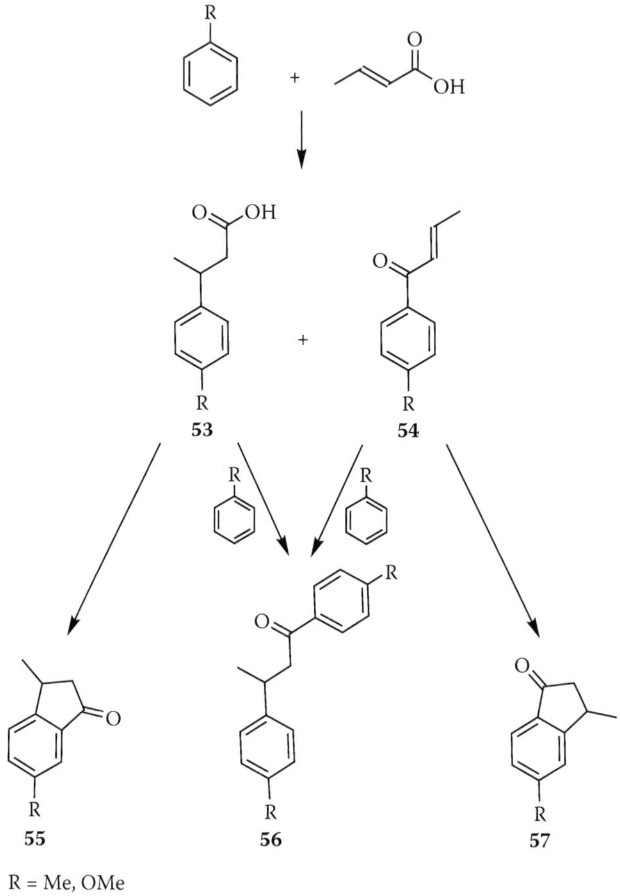

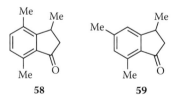

Scheme 4.36

Figure 4.11 Product obtained in the intramolecular acylation/alkylation of xylenes and crotonic acid catalyzed by HPA/SiO$_2$.

anions are unable to stabilize such cations. However, it is important to take into account that a similar behavior is observed with acid-treated metal oxides.[168,173] No dissolution of catalytically active species from the salts is observed (the reaction ceases completely if the heterogeneous salt is removed from the reactor). $Rb_{2.5}H_{0.5}[PW_{12}O_{40}]$ and $Cs_{2.5}H_{0.5}[PW_{12}O_{40}]$ (CsPW) are highly porous materials with mesopores showing 2–7 nm average diameter; this mesoporous structure seems to favor benzoylation since the reaction takes place not only on the surface but also in the bulk near the surface owing to the highly polar nature of BC and the deriving cationic intermediate.

Anhydrides are also employed in the acylation of aromatics with ion-exchanged HPAs. For example, the insoluble CsPW can be applied in the acylation of benzene with BAN[201]; BP is obtained in 100% yield. The initial rate is almost proportional to the amount of catalyst. Repeated use of the catalyst was possible by washing the used one with benzene. However, the activity gradually decreases after the third run due to the adsorption of both BP and BAC.

More interestingly, carboxylic acids can be employed in the acylation of aromatic compounds. The acylation of toluene and anisole with C_2–C_{12} aliphatic carboxylic acids can be carried out with CsPW, affording the corresponding products in 41%–71% yield.[202] These solid acids are superior in activity to the conventional acid catalysts such as sulfuric acid and zeolites, and can be reused after a simple workup, albeit with reduced activity.

The activity of the catalyst can be improved by carrying out the reaction in the presence of TFAA.[203] The acylation of anisole is conducted by acetic or benzoic acid in the presence of TFAA catalyzed by $AlPW_{12}O_{40}$ with excellent yields (94%–96%). Under the same reaction conditions, 2-methyl and 4-methyl anisoles are also acylated, giving the corresponding ketones in 90%–97% yield.

As already underlined by Kozhevnikov,[185,186] it is quite evident that HPAs and related compounds represent materials to be efficiently utilized as catalysts in organic reactions, including Friedel–Crafts acylation. However, a serious problem with the use of these compounds as heterogeneous catalysts is their deactivation because of coking. In fact, conventional regeneration by calcining at 500°C–550°C, which is routinely used in the case of oxides and zeolites, cannot be applied to HPAs because their thermal stability is not enough. However, some tricks useful in alkane isomerization processes, which include doping with transition metal ions (i.e., Pd) and controlled addition of water to the catalyst, can be successfully utilized to prolong the lifetime of HPA catalysts.

CsPW is a better catalyst than the PW itself, but its particles are very fine, and separation of the catalyst from the liquid remains a problem. For this reason, the salt is supported on K10 montmorillonite. The catalyst

CsPWK10 with intact Keggin anion structure gives satisfactory results in the benzoylation of *para*-xylene with BC, affording 2,5-dimethylbenzophenone in 60% yield.

Similar good results can be obtained in the benzoylation of anisole with BC. At 110°C, an 80% conversion of the acyl chloride is achieved after 2 h with CsPWK10 (20% loading), *para*-methoxybenzophenone being the sole product obtained. The catalyst can be efficiently reused three times.[204]

4.6 Nafion

Nafion (Figure 4.12), a perfluorinated polymer containing pendant sulfonic acid groups, is generally considered to be a solid superacid whose pK_a ranges from −5 to −9. It was earlier shown to be an active catalyst for various organic reactions such as alkylation, isomerization, disproportionation, transalkylation, acylation, nitration, hydration, rearrangement, and so on.[205,206]

The use of Nafion in Friedel–Crafts acylation reactions was reported[207] for the benzoylation of toluene with BCs. The reaction can be simply carried out by heating under reflux the mixture of the reagents with the solid catalyst, affording substituted BPs in 81%–87% yield. Higher catalyst quantities significantly decrease the yields due to adsorption of appreciable amounts of products and starting materials. Concerning the isomeric composition of methylbenzophenones, the reaction gives predominantly para and ortho substitution (in accordance with the typical electrophilic aromatic reaction), but the ortho/para ratio (2.3 4.8) is, in general, higher than that obtained under homogenous (aluminum chloride) catalysis (0.1–0.6). The reaction can be applied to *para*-nitrobenzoyl chloride and various substituted benzenes, giving the corresponding BPs in 60%–90% yield.

More problematic results are obtained with the acetylation of aromatic compounds; in fact, with AC, the preferential formation of ketene degradation products is observed, whereas neither AAN nor AAC alone prove to be effective. On the contrary, when an equimolecular mixture of AAC and AAN is refluxed with toluene, *meta*-xylene, or mesitylene in the presence of Nafion, the corresponding APs are isolated in 3%, 21%, and 72% yield, respectively.

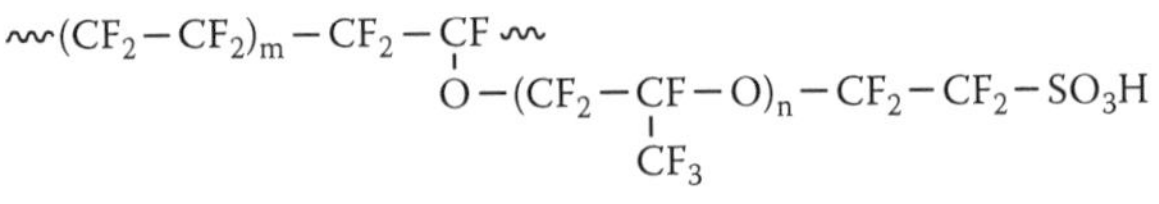

$$\sim\!\!(CF_2-CF_2)_m - CF_2 - \underset{\underset{\displaystyle O-(CF_2-\underset{\underset{CF_3}{|}}{CF}-O)_n - CF_2 - CF_2 - SO_3H}{|}}{CF}\!\!\sim$$

Figure 4.12 Nafion catalyst structure.

Table 4.23 Preparation of fluorenone and related cyclic and heterocycles ketones via cycloacylation of the corresponding aryl carboxylic acids promoted by Nafion

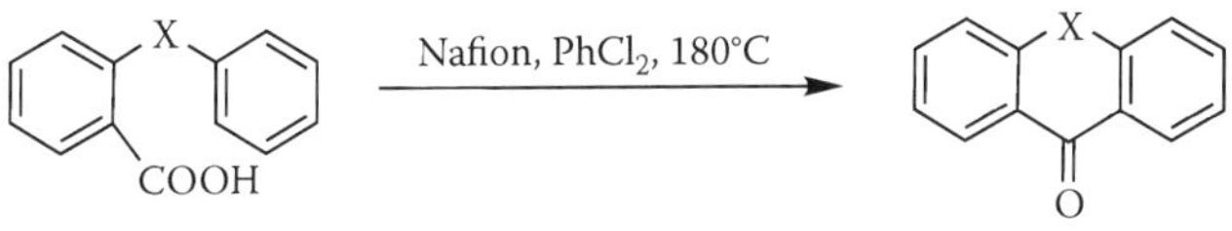

X	t (h)	Yield (%)
bond	3	82
C=O	3	90
CH_2	2	92
NH	2	85
O	1	90
CH_2CH_2	3	95

Preparation of fluorenone and related cyclic and heterocyclic ketones **61** from the corresponding aryl carboxylic acids **60** or the appropriate BAC derivatives can be performed under relatively mild conditions (Table 4.23).[208,209] The reaction is carried out by heating a mixture of the carboxylic acid derivative **60** and the solid Nafion in 1,2-dichlorobenzene at about 180°C, affording products **61** in excellent yields. The reaction can be also performed in refluxing *para*-xylene in the case of the carboxylic acid (100% yield with diphenylethane-2-carboxylic acid after 12 h) and in refluxing benzene with the corresponding acyl chloride, giving similar good results.

Moreover, 4-arylbutanoic acids **62** cyclize in refluxing *para*-xylene in the presence of Nafion, giving 1-tetralone analogs **63** in nearly quantitative yields (Scheme 4.37).

The cyclization of 3-phenylpropanoyl acid chloride to 1-indanone can be performed in boiling benzene with 90% yield. The in situ formation of mixed carboxylic–fluorosulfonic anhydride intermediates is suggested to account for the high catalytic activity of Nafion.

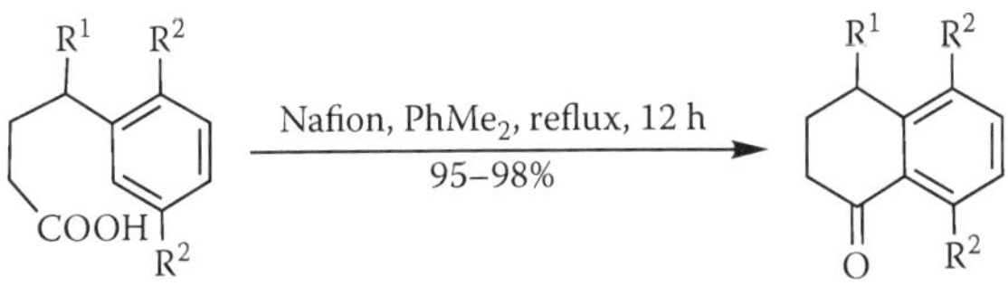

Scheme 4.37

Nafion silica composites are successfully utilized in the acylation of aromatics with both carboxylic acids and acyl chlorides. The composite materials are prepared by entrapping highly dispersed nanosized Nafion particles within a silica matrix.[210] These catalysts show the advantages of both components, namely, the acidic strength of Nafion resin and the simple accessibility of the high surface area of silica. This combination results in high catalytic activity at both moderate temperature and low catalyst loadings, which often guarantees high selectivity to the desired product. By using a composite catalyst containing 13% nanosized Nafion particles, anisole is acylated with phenylacetyl chloride to give 2-phenyl-(4′-methoxy)acetophenone. It was observed that the composite material gives an increase of acyl chloride conversion from 71% to 100% by increasing the temperature from 60°C to 100°C, whereas the pure unsupported Nafion achieves 46% to 77% conversion during the same rise in temperature. Moreover, the selectivity to the desired *para*-ketone steadily rises (from 81% to 85%) as the temperature increases from 60°C to 100°C. This apparently anomalous trend can be explained taking into account that the most abundant by-products are the phenylacetic acid and the vinyl chloride derivative **64** (Scheme 4.38) already observed in the presence of hard acid catalysts.[160] By enhancing the reaction temperature, the solubility of the hydrogen chloride in the reaction medium is distinctly reduced; consequently, selectivity to the desired product is enhanced.

In studies on the acylation of anisole with octanoic acid, it is observed that carboxylic acids, when present in large amounts, adsorb so strongly on the catalyst surface that the aromatic substrates have no chance to react.[211]

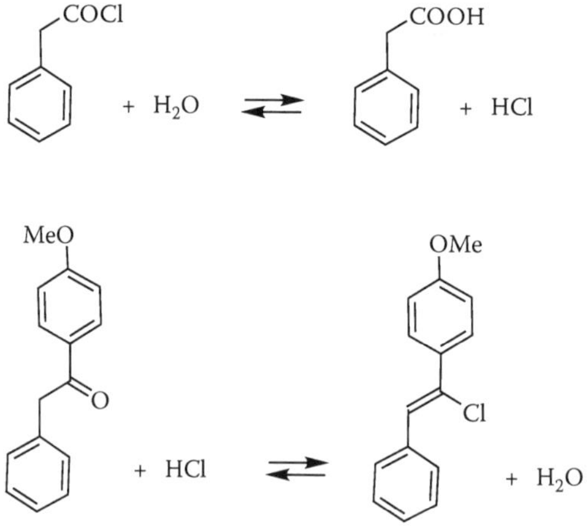

Scheme 4.38

Consequently, a continuous process can be preferred since it operates at low acid concentrations. This is achievable by using a fixed-bed catalyst with a recirculation of the reaction mixture. A structured catalyst is therefore needed to allow for higher recirculation rates and to avoid a high pressure drop over the catalyst. Nafion/silica composite materials, which are more active in the acylation of anisole with octanoic acid than BEA and USY zeolites and Amberlyst-15 resin, are good structured catalysts for this purpose. The highest activity is exhibited by the Nafion/silica composite containing more accessible Nafion nanoparticles even if a modest para-selectivity is obtained, ranging from 44% to 72% and being the major by-product the O-acylated compound. Removal of water produced during the reaction has a beneficial effect, and a fourfold growing of catalyst activity is observed.

The use of coated monoliths as supports was further studied; the catalyst is present as a thin layer on the channel walls. The advantages of such a system are that no filtering of catalyst is necessary, no attrition of catalyst occurs, and the reactor has a low pressure drop; in addition, the reaction is performed in the liquid phase, and the water is removed through the gas phase. Two different types of cordierite monoliths are employed: the first with 400 cells per square inch (cpsi) and 1.05 wt% Nafion loading, and the second with 600 cpsi and 1.27 wt% Nafion loading; also, a 400 cpsi silica-coated monolith with 0.96 wt% Nafion loading is tested. The 400 and 600 cpsi monoliths exhibit the same selectivity (46%–47%) as the pure Nafion in batch reaction (47%), whereas the silica-coated 400 cpsi monolith shows a higher selectivity (70%); these results indicate that the silica matrix leads to higher selectivities, probably due to the interaction of Nafion with the silica surface.

Perfluoroalkylsulfonic acid chains can be anchored to MCM-41 silica.[212] The supported catalyst is prepared by reacting MCM-41 silica with 1,2,2-trifluoro-2-hydroxy-1-trifluoromethylethane sulfonic acid beta-sultone **65** (Scheme 4.39). With this catalyst, stable up to 350°C, the acylation of anisole with AAN is carried out, showing a good anisole conversion (35%–45%) and a very high selectivity (97%) for the desired para-isomer at 60°C. The best results in terms of conversion are achieved with the

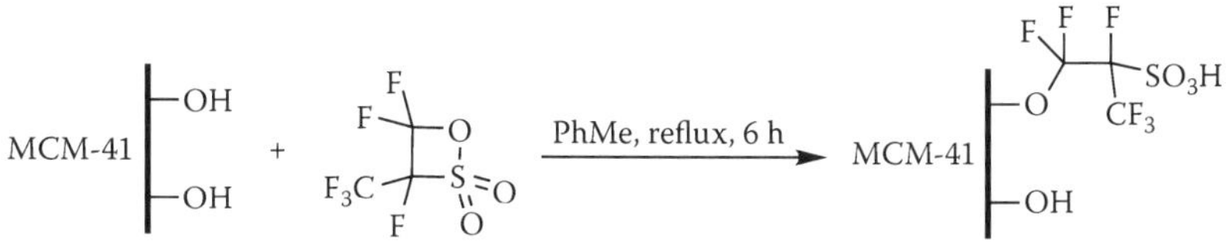

Scheme 4.39

catalyst showing the highest loading; an increase in reaction temperature up to 120°C allows isolation of the product in ~60% yield. A limit on the exploitation of such a catalyst is represented by its deactivation; further studies are needed to establish the stability of the covalent bonds between the alkyl sulfonic groups and silica surface to the hydrolysis under strong acid conditions and the influence of the residual silanols on the acid strength of the perfluoroalkylsulfonic groups.

A series of silica-surface-bounded strong solid acid catalysts can be prepared by using $(OH)_3Si(CH_2)_3(CF_2)_2O(CF_2)_2SO_3^-K^+$.[218,219] This compound can be hydrolized and condensed with TEOS, giving very-high-surface-area materials with pendant perfluorosulfonic acid groups, whose surface area, pore volume, and pore diameter are found to be 550 m^2 × g^{-1}, 0.75 cm^3 × g^{-1}, and 55 Å, respectively. Alternatively, it can be anchored to high-surface-area silica support (~400 m^2 × g^{-1}), according to the tethering approach.[220] Both materials show high activity in the benzoylation of *ortho*-xylene with BC at 140°C for 6 h, giving a BC conversion of ~90%.

A detailed analytical study of the activity of some solid acid catalysts, including mesoporous silica-supported Nafion, in the acylation of anisole with AAN allows the conclusion that catalyst deactivation is caused by the primary ketone product and/or multiple acetylated products in the micropores of Nafion catalyst aggregates.[213] Experiments were performed with a commercially available silica-supported Nafion catalyst in a continuous-mode slurry operation by using carbon-dioxide-expanded liquids (nitromethane or nitrobenzene) as solvents. At 90°C, 80% AAN conversion is observed with a TOS of 2 h, but the catalyst rapidly deactivates, and 27% conversion is evaluated after 6 h TOS with a TON value of about 400. The catalyst can, however, be completely regenerated upon nitric acid treatment. These results confirm that silica-supported Nafion catalysts are promising alternatives for the traditional aluminum chloride homogeneous Lewis acid catalyst.

4.7 Miscellaneous

Graphite, without any treatment, can promote the Friedel–Crafts acylation of aromatic compounds with acyl halides to afford the corresponding acylated products.[214] This material has a remarkably high activity for the acylation of active aromatic compounds such as anisoles and polymethylbenzenes with acyl halides. In a typical experiment, graphite is added to a mixture of anisole and benzoyl bromide in benzene, and the mixture is heated under reflux for 8 h to afford *para*-methoxybenzophenone in 80% yield. When a large amount of graphite (1 g per 2 mmol of anisole) is used, the yields are lower because of absorption of an appreciable quantity of the product and starting material on the graphite. Different acyl halides

Table 4.24 Aromatic acylation with acyl halides catalyzed by graphite

R^1	R^2	R^3	R^4	R^5	R^6	X	t (h)	Yield (%)
H	H	OMe	H	H	Ph	Br	8	89
H	OMe	OMe	H	H	Ph	Br	8	92
OMe	H	H	OMe	H	Ph	Br	8	60
H	H	OMe	CH=CH-CH=CH		Ph	Br	8	93
H	Me	Me	H	H	Ph	Br	24	97
Me	H	Me	H	Me	Ph	Br	3	84
Me	Me	Me	H	Me	Ph	Br	3	85
Me	Me	Me	Me	Me	Ph	Br	3	83
H	H	OMe	H	H	Ph	Cl	8	85
H	H	OMe	H	H	c-C_6H_{11}	Cl	8	84
H	H	OMe	H	H	Pr^i	Cl	3	89
H	H	OMe	H	H	C_5H_{11}	Cl	8	66
H	H	OMe	H	H	$C_{12}H_{25}$	Cl	8	63
H	H	OMe	H	H	Bn	Cl	8	61

and several anisoles and polymethylbenzenes can be reacted to obtain the corresponding products in high yields (Table 4.24).

To overcome the problem of the high amount of graphite utilized and mainly to avoid the use of the expensive and pollutant acyl halides and benzene, graphite can be coupled with *para*-toluenesulfonic acid and utilized to activate the more ecocompatible carboxylic acids toward electrophilic acylation.[215] With this catalyst, not only a solvent-free process can be developed but also high yields with not activated aromatic substrates and with unactivated carboxylic acids (i.e., *para*-nitrobenzoic acid) can be obtained (Table 4.25).

Interestingly, in the presence of graphite or *para*-toluenesulfonic acid alone, no acylated products are isolated. The effect of the solvent is detrimental since only traces of the products are detected when the reactions are carried out in methylene chloride or chloroform. It must be underlined that graphite can be reused after simple washing with ethyl acetate and water, but the *para*-toluenesulfonic acid, which is not adsorbed on the graphite during the reaction, must be added again for the successive runs.

The sol–gel procedure is utilized to immobilize various triflate catalysts in a silica matrix.[216] In particular, *tert*-butyldimethylsilyltriflate (BDMST)

Table 4.25 Aromatic acylation with carboxylic acids in the presence of graphite–*para*-toluenesulfonic acid mixture

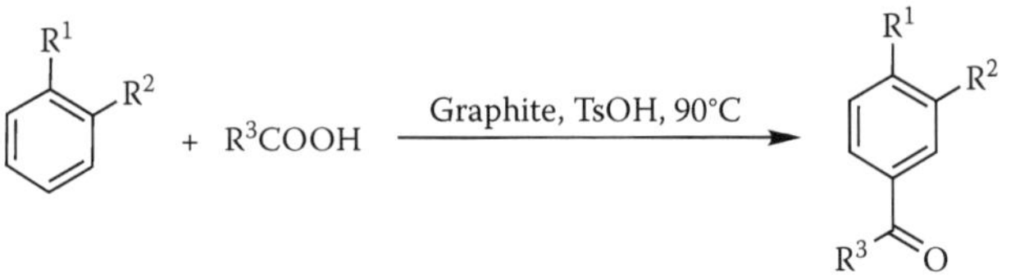

R^1	R^2	R^3	t (h)	Yield (%)
OMe	H	Me	3	96
OMe	OMe	Me	6	87
Me	H	Me	6	85
Me	Me	Me	5	81
OMe	H	2-pyridyl	2.5	90
OMe	H	3-ClC$_6$H$_4$	1.5	95
OMe	H	4-NO$_2$C$_6$H$_4$	4	90
OMe	H	2-OHC$_6$H$_4$	3.5	85
OMe	H	c-C$_6$H$_{11}$	1	90
OMe	H	C$_9$H$_{19}$	3	92
OMe	H	Ph	3	86
OMe	H	BrC$_{10}$H$_{20}$	4	84
OMe	H	Bn	3.5	92

can be immobilized on silica by acid hydrolysis of a water solution of TEOS, giving a sol to which BDMST is added; the gelation is continued for 2 days.[217] The acylation of 2-methoxynaphtahlene with AAN in the presence of the just-described prepared catalyst is examined. 1-Acetyl-2-MN is the major product accompanied by trace amounts of 2-naphthol. The best result (89% conversion and 98% selectivity) is achieved by performing the reaction without solvent at 50°C. Similar good yield and selectivity can be achieved with silica-supported triflic acid under the same conditions.

Detailed kinetic studies confirm that the reaction with BDMST-supported catalyst largely occurs in the first 10 min and the TOF value amounts to 208 min⁻¹, which is remarkable in comparison with the values typically recorded with zeolites.

Some information is available on the solvent effect; better results are obtained in ethers than in alcohols. The overall order is solventless > 1,4-dioxane > tetrahydrofuran > ethanol ~ isopropanol. The polarity of the solvent is an important parameter affecting acylation because polar solvents compete more successfully with the reagents for adsorption and diffusion on the pores of the catalyst. On the other hand, they provide a better solvation for the acylium intermediate. A good relation is found

Table 4.26 Correlation between the solvent DN and the conversion of 2-MN in the acylation with AAN under BDMST catalysis at 50°C for 4 h

Solvent	DN (KJ/mol)	Conversion (%)	Product distribution (%)		
			1-Ac-2-MN	2-Naphthol	Others
—	43.9	85	97.8	1.7	0.5
1,4-Dioxane	61.9	25	98.7	1.3	—
THF	83.7	5	99	<1	—
EtOH	133.9	<1	—	—	—
$Pr^{i}OH$	150.7	<1	—	—	—

between the reaction rate and the donor number (DN) of the solvent: the higher the DN value, the lower the reaction rate (Table 4.26).

The bentonite-supported polytrifluoromethanesulfosiloxane (B-PTF-MSS) shows good catalytic activity in the electrophilic acylation of ferrocene with different acyl chlorides to afford products **66** (Scheme 4.40).[221] The catalyst B-PTFMSS can be prepared by refluxing a stirred suspension of bentonite and triflc acid in ethanol and calcining at 200°C for 4 h. The maximum efficiency achieved is ~80% ferrocene conversion. The catalyst can be reused with similar activity for two further cycles. Similar good results are achieved with aliphatic acyl chlorides (~60%–75% yield of product **66**), whereas poor yields are obtained with chlorobutyryl chloride (18%), ferrocenoyl chloride (8%), and BC (8%).

Triflic-acid-functionalized mesoporous zirconias (Tf-ZrO_2) can be prepared by postsynthesis functionalization of mesoporous zirconia with triflic acid in refluxing toluene.[222,223] Mesoporous zirconia is, in turn, prepared by the hydrolysis of zirconium(IV) butoxide to zirconium(IV) hydroxide and stirring the precipitate with *N*-acetyl-*N*,*N*,*N*-trimethylammonium bromide and tetramethylammonium hydroxide. Catalysts containing increasing amounts of anchored triflic acid (from 5% to 30% mol) give interesting results in the benzoylation of toluene with *para*-toluoyl chloride **67** for the production of 4,4′-dimethylbenzophenone **68** (Table 4.27).[224,225]

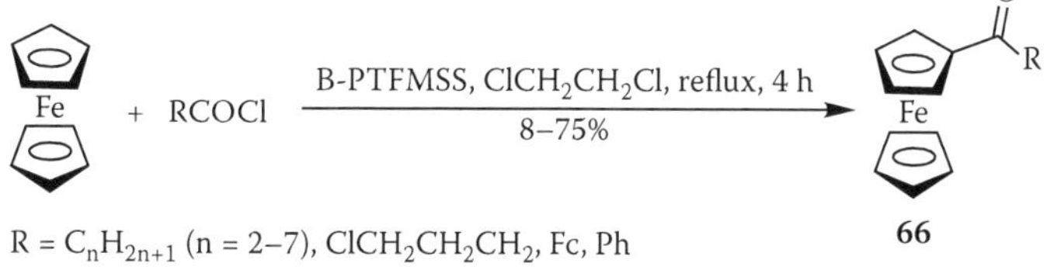
Fe
+ RCOCl
B-PTFMSS, $ClCH_2CH_2Cl$, reflux, 4 h
8–75%
O
R
Fe
$R = C_nH_{2n+1}$ (n = 2–7), $ClCH_2CH_2CH_2$, Fc, Ph

66

Scheme 4.40

Table 4.27 Toluene benzoylation with *para*-toluoyl chloride over different acid catalysts

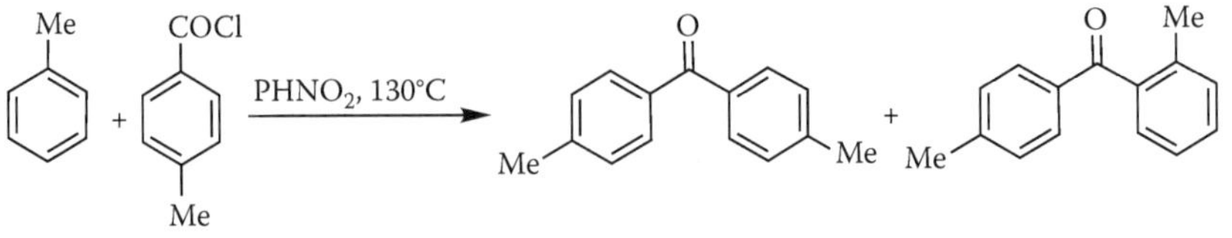

Catalyst	t (h)	TOF (h⁻¹)	67 conversion (%)	Product distribution (%)		
				68	69	Others
Zr₂O	8	6.4	22	22.0	78.0	0.0
Zr₂O	24	4.0	42	15.5	78.6	5.9
Tf-Zr₂O-5	8	13.2	46	22.1	75.9	2.0
Tf-Zr₂O-5	24	6.8	72	23.0	73.3	3.7
Tf-Zr₂O-15	8	14.5	51	22.5	73.5	4.0
Tf-Zr₂O-15	24	7.8	82	21.4	74.1	4.5
Tf-Zr₂O-30	8	15.5	54	22.5	74.2	3.3
Tf-Zr₂O-30	24	8.1	85	22.8	73.9	3.3
TfOH	0.25	420.1	88	20.1	76.4	3.5
TfOH	3	40.0	100	21.2	75.2	3.6

The reaction also produces 2,4′-dimethylbenzophenone **69**; the conversion of **67** as well as the isomer molar ratio depend on the type of catalyst used. Among the Tf-ZrO₂ with various loadings of triflic acid, Tf-ZrO₂-30 reveals the highest conversion of **67**, which may be attributed to its stronger acid sites, as is also confirmed by ammonia desorption experiments. The conversion of **67** increases significantly with an increase in reaction time, catalyst concentration, reaction temperature, and toluene/**67** molar ratio. The catalyst can be recycled one time, and some decrease in **67** conversion is observed (51% fresh, 49% 1st recycle), which is related to the minor leaching of anchored triflic acid by the hydrogen chloride formed during the reaction.

As sulfur-containing organic molecules are known catalyst poisons because of strong adsorption, acylation of thioethers is difficult to obtain. The reaction between thioanisole and AAN can, however, be performed in the presence of ion-exchange resins that are more robust to deactivation.[226,227] The process is carried out in a Parr autoclave in 1,2-dichloroethane at 70°C. Under these conditions, other acid catalysts such as sulfated zirconia and K10 clay do not show any noticeable activity. Only the cation exchange resin catalysts, which contain Brönsted sites, are effective. Among these, Amberlyst-15 shows maximum conversion because it has bigger pores and higher surface area; the selectivity to the desired product at the end of the experiment is over 98%.

Sulfonic acid supported on mesostructured SBA-15 silica is studied as a catalyst in the acylation of anisole and 2-MN with AAN.[228] The catalyst is prepared by hydrolyzing a mixture of TEOS and 3-mercaptopropyl trimethoxysilane in the presence of hydrogen peroxide. This catalyst shows a greater activity as compared to other homogeneous and heterogeneous sulfonated materials. The selectivity toward *para*-methoxyacetophenone is higher than 95%, showing that these results come from specific anisole reactivity rather than the shape-selectivity effect. The supported catalysts display the highest activity (TON = 85–93) with respect to both heterogeneous Amberlyst-15 (TON = 40) and homogeneous *para*-toluenesulfonic acid (TON = 5) catalysts. The significant decrease of conversion rate per acid center observed with Amberlyst-15 can be related to the low accessibility of the acid sites arising from its low specific surface area ($45 \text{ m}^2 \times \text{g}^{-1}$).

An efficient MCM-41-based catalyst is set up by impregnating sulfuric acid on AlMCM-41 (SO_4^{2-}/AlMCM-41);[229] this composite material shows a good catalytic efficiency in the preparation of 2-acetyl-6-MN by reaction of 2-MN with AAN, giving 60% conversion and 43% yield.

References

1. Corma, A. and Garcia, E. 1997. Organic reactions catalyzed over solid acids. *Catal. Today* 38: 257–308.
2. Čejka, J. and Wichterlová, B. 2002. Acid-catalyzed synthesis of mono- and dialkyl benzenes over zeolites: active sites, zeolite topology, and reaction mechanisms. *Catal. Rev.* 44: 375–421.
3. Meier, W. M. and Olson, D. H. 1992. Atlas of zeolite structure types. *Zeolites* 12: 449–656.
4. Wilson, S. T., Lok, B. M., Messina, C. A., Cannan, T. R., and Flanigen, E. M. 1982. Aluminophosphate molecular sieves: a new class of microporous crystalline inorganic solids. *J. Am. Chem. Soc.* 104: 1146–1147.
5. Lok, B. M., Messina, C. A., Patton, R. L., Gajek, R. T., Cannan, T. R., and Flanigen, E. M. 1984. Silicoaluminophosphate molecular sieves: another new class of microporous crystalline inorganic solids. *J. Am. Chem. Soc.* 106: 6092–6093.
6. Corma, A. 1995. Inorganic solid acids and their use in acid-catalyzed hydrocarbon reactions. *Chem. Rev.* 95: 559–614.
7. Beck, J. S., Vartuli, J. C., Roth, W. J., Leonowicz, M. E., Kresge, C. T., Schmitt, K. D., Chu, W., Olson, D. H., Sheppard, E. W., McCullen, S. B., Higgins, J. B., and Schlenker, J. L. 1992. A new family of mesoporous molecular sieves prepared with liquid crystal templates. *J. Am. Chem. Soc.* 114: 10834–10843.
8. Klisáková, J., Cervený, L., and Čejka, J. 2004. On the role of zeolite structure and acidity in toluene acylation with isobutyric acid derivatives. *Appl. Catal. A: Gen.* 272: 79–86.

9. Moreau, P., Finiels, A., Pelorgeas, S., Vigneau, O., and Lasperas, M. 1997. Liquid-phase acetylation of tetralin with acetyl chloride over zeolites: inhibition of the reaction by the products. *Catal. Lett.* 47: 161–166.

10. Chiche, B., Finiels, A., Gauthier, C., and Geneste, P. 1986. Friedel–Crafts acylation of toluene and *p*-xylene with carboxylic acids catalyzed by zeolites. *J. Org. Chem.* 51: 2128–2130.

11. Isaev, Y. and Fripiat, J. J. 1999. A Lewis acid site-activated reaction in zeolites: thiophene acylation by butyryl chloride. *J. Catal.* 182: 257–263.

12. Fang, R., Harvey, G., Kouwenhoven, H. W., and Prins, R. 1995. Effects of non-framework alumina in the acylation of xylene over USY. *Appl. Catal. A: Gen.* 130: 67–77.

13. Fang, R., Kouwenhoven, H. W., and Prins, R. 1994. Benzoylation of xylenes using zeolitic catalysts. *Stud. Surf. Sci. Catal.* 84: 1441–1448.

14. Coster, D., Blumenfeld, A. L., and Fripiat, J. J. 1994. Lewis acid sites and surface aluminum in aluminas and zeolites: a high-resolution NMR study. *J. Phys. Chem.* 98: 6201–6211.

15. Olah, G. A. 1973. *Friedel–Crafts chemistry*. New York: Wiley-Interscience.

16. Spagnol, M., Gilbert, L., Benazzi, E., and Marcilly, C. 2000. Process for the acylation of aromatic ethers. U.S. Patent 6,013,840.

17. Gauthier, C., Chiche, B., Finiels, A., and Geneste, P. 1989. Influence of acidity in Friedel–Crafts acylation catalyzed by zeolites. *J. Mol. Catal.* 50: 219–229.

18. Chiche, B., Finiels, A., Gauthier, C., and Geneste, P. 1987. The effect of structure on reactivity in zeolite catalyzed acylation of aromatic compounds: a $\rho^- \sigma^+$ relationship. *Appl. Catal.* 30: 365–369.

19. Harbison, K. G. 1970. Isomerization of xylenes. Experiment for the organic or instrumental laboratory. *J. Chem. Educ.* 47: 837–839.

20. Chiche, B. H., Gauthier, C., and Geneste, P. 1987. Process for the acylation of aromatic hydrocarbons. FR Patent 2,592,039.

21. Badri, R. and Tavakoli, L. 2003. Friedel–Crafts acylation of aryl-substituted carboxylic acids catalyzed by zeolite. *J. Inclusion Phenom. Macrocyclic Chem.* 45: 41–43.

22. Gaare, K. and Akporiaye, D. 1996. Modified zeolites as catalysts in the Friedel–Crafts acylation. *J. Mol. Catal. A: Chem.* 109: 177–187.

23. Sreekumar, R., Rugmii, P., and Padmakumar, R. 1998. Synthesis of azafluorenones using zeolites. *Synth. Commun.* 28: 2071–2075.

24. Laidlaw, P., Bethell, D., Brown, S. M., and Hutchings, G. J. 2001. Benzoylation of substituted arenes using Zn- and Fe-exchanged zeolites as catalysts. *J. Mol. Catal. A: Chem.* 174: 187–191.

25. Ma, Y., Wang, Q. L., Jiang, W., and Zuo, B. 1997. Friedel–Crafts acylation of anisole over zeolite catalysts. *Appl. Catal. A: Gen.* 165: 199–206.

26. Corma, A., Climent, M. J., García, E., and Primo, J. 1989. Design of synthetic zeolites as catalysts in organic reactions. Acylation of anisole by acyl chlorides or carboxylic acids over acid zeolites. *Appl. Catal.* 49: 109–123.

27. Akporiaye, D. E., Daasvatn, K., Solberg, J., and Stöcker, M. 1993. Modified zeolites as active catalysts in Friedel–Crafts acylation. *Stud. Surf. Sci. Catal.* 78: 521–526.

28. Hardacre, C., Katdare, S. P., Milroy, D., Nancarrow, P., Rooney, D. W., and Thompson, J. M. 2004. A catalytic and mechanistic study of the Friedel–Crafts benzoylation of anisole using zeolites in ionic liquids. *J. Catal.* 227: 44–52.

29. Smith, K., El-Hiti, G. A., Jayne, A. J., and Butters, M. 2003. Acetylation of aromatic ethers using acetic anhydride over solid acid catalysts in a solvent-free system: scope of the reaction for substituted ethers. *Org. Biomol. Chem.* 1: 1560–1564.

30. Bigi, F., Carloni, S., Flego, C., Maggi, R., Mazzacani, A., Rastelli, M., and Sartori, G. 2002. HY zeolite-promoted electrophilic acylation of methoxyarenes with linear acid chlorides. *J. Mol. Catal. A: Chem.* 178: 139–146.

31. Raja, T., Singh, A. P., Ramaswamy, A. V., Finiels, A., and Moreau, P. 2001. Benzoylation of 1,2-dimethoxybenzene with benzoic anhydride and substituted benzoyl chlorides over large pore zeolites. *Appl. Catal. A: Gen.* 211: 31–39.

32. Moreau, P., Finiels, A., and Meric, P. 2000. Acetylation of dimethoxybenzenes with acetic anhydride in the presence of acidic zeolites. *J. Mol. Catal. A: Chem.* 154: 185–192.

33. Spagnol, M., Gilbert, L., Benazzi, E., and Marcilly, C. 2000. Process for the acylation of aromatic ethers. US patent 6,013,840.

34. Richard, F., Drouillard, J., Carreyre, H., Lemberton, J. L., and Pérot, G. 1993. Zeolite catalyzed acylation of heterocyclic aromatic compounds. I. Acylation of benzofuran. *Stud. Surf. Sci. Catal.* 78: 601–606.

35. Richard, F., Carreyre, H., and Pérot, G. 1995. Zeolite catalyzed acylation of heterocyclic compounds. Part III. Comparison between benzofuran and 2-methylbenzofuran. *J. Mol. Catal. A: Chem.* 101: L167–L169.

36. Richard, F., Carreyre, H., and Pérot, G. 1995. Zeolite catalyzed acylation of heterocyclic compounds. Part II. Acylation of benzofuran over Y zeolites. Effect of reaction conditions on the activity and stability. *J. Mol. Catal. A: Chem.* 103: 51–61.

37. Richard, F., Carreyre, H., and Pérot, G. 1996. Zeolite-catalyzed acylation of heterocyclic compounds: acylation of benzofuran and 2-methylbenzofuran in a fixed bed reactor. *J. Catal.* 159: 427–434.

38. Amouzegh, P., Finiels, A., Geneste, P., Ginestar, E., and Moreau, P. 1995. Acylation of a substituted benzofuran over an HY zeolite and its subsequent deacylation and reacylation. *Catal. Lett.* 34: 389–394.

39. Finiels, A., Calamettes, A., Geneste, P., and Moreau, P. 1993. Kinetic study of the acylation of thiophene with acyl chlorides in liquid phase over HY zeolites. *Stud. Surf. Sci. Catal.* 78: 595–600.

40. Botella, P., Corma, A., López-Nieto, J. M., Valencia, S., and Jacquot, R. 2000. Acylation of toluene with acetic anhydride over Beta zeolites: influence of reaction conditions and physicochemical properties of the catalyst. *J. Catal.* 195: 161–168.

41. Singh, A. P. and Bhattacharya, D. 1995. Benzoylation of benzene to benzophenone over zeolite catalysts. *Catal. Lett.* 32: 327–333.

42. Singh, A. P., Bhattacharya, D., and Sharma, S. 1995. Benzoylation of toluene with benzoyl chloride over zeolite catalysts. *J. Mol. Catal. A: Chem.* 102: 139–145.

43. Venkatesan, C., Jaimol, T., Moreau, P., Finiels, A., Ramaswamy, A. V., and Singh, A. P. 2001. Liquid phase selective benzoylation of chlorobenzene to 4,4′-dichlorobenzophenone over zeolite H-Beta. *Catal. Lett.* 75: 119–123.

44. Jacob, B., Sugunan, S., and Singh, A. P. 1999. Selective benzoylation of *o*-xylene to 3,4-dimethylbenzophenone using various zeolite catalysts. *J. Mol. Catal. A: Chem.* 139: 43–53.

45. Santacesaria, E., Scaglione, A., Apicella, B., Tesser, R., and Di Serio, M. 2001. Synthesis and purification of anthraquinone in a multifunctional reactor. *Catal. Today* 66: 167–174.

46. Kikhtyanin, O. V., Ione, K. G., Snytnikova, G. P., Malysheva, L. V., Toktarev, A. V., Paukshtis, E. A., Spichtinger, R., Schüth, F., and Unger, K. K. 1994. Anthraquinones formation on zeolites with BEA structure. *Stud. Surf. Sci. Catal.* 84: 1905–1912.

47. Bourgogne, J.-P., Aspisi, C., Ou, K., Geneste, P., Durand, R., and Mseddi, S. 1992. Process for the acylation of aromatic hydrocarbons. FR Patent 2,667,063.

48. Chidambaram, M., Venkatesan, C., Moreaub, P., Finiels, A., Ramaswamy, A. V., and Singh, A. P. 2002. Selective benzoylation of biphenyl to 4-phenyl-benzophenone over zeolite H-Beta. *Appl. Catal. A: Gen.* 224: 129–140.

49. Escola, J. M. and Davis, M. E. 2001. Acylation of biphenyl with acetic anhydride and carboxylic acids over zeolite catalysts. *Appl. Catal. A: Gen.* 214: 111–120.

50. Choudhary, V. R., Jana, S. K., Patil, N. S., and Bhargava, S. K. 2003. Friedel–Crafts type benzylation and benzoylation of aromatic compounds over Hβ zeolite modified by oxides or chlorides of gallium and indium. *Microporous Mesoporous Mater.* 57: 21–35.

51. Sheemol, V. N., Tyagi, B., and Jasra, R. V. 2004. Acylation of toluene using rare earth cation exchanged zeolite β as solid acid catalyst. *J. Mol. Catal. A: Chem.* 215: 201–208.

52. Choudary, B. M., Sateesh, M., Kantam, M. L., Kururi, B. S. R., and Raghavan, K. V. 2001. Method of preparing 4′-isobutyl acetophenone. JP Patent 2,001,278,833.

53. Červený, L., Mikulcová, K., and Čejka, J. 2002. Shape-selective synthesis of 2-acetylnaphthalene via naphthalene acylation with acetic anhydride over large pore zeolites. *Appl. Catal. A: Gen.* 223: 65–72.

54. Bhattacharya, D., Sharma, S., and Singh, A. P. 1997. Selective benzoylation of naphthalene to 2-benzoylnaphthalene using zeolite H-Beta catalysts. *Appl. Catal. A: Gen.* 150: 53–62.

55. Derouane, E. G., Dillon, C. J., Bethell, D., and Derouane-Abd Hamid, S. B. 1999. Zeolite catalysts as solid solvents in fine chemicals synthesis. 1. Catalyst deactivation in the Friedel–Crafts acetylation of anisole. *J. Catal.* 187: 209–218.

56. Derouane, E. G., Crehan, G., Dillon, C. J., Bethell, D., He, H., and Derouane-Abd Hamid, S. B. 2000. Zeolite catalysts as solid solvents in fine chemicals synthesis. 2. Competitive adsorption of the reactants and products in the Friedel-Crafts acetylations of anisole and toluene. *J. Catal.* 194: 410–423.

57. Derouane, E. G. 1998. Zeolites as solid solvents. *J. Mol. Catal. A: Chem.* 134: 29–45.

58. Derouane, E. G., Schmidt, I., Lachas, H., and Christensen, C. J. H. 2004. Improved performance of nano-size H-BEA zeolite catalysts for the Friedel-Crafts acetylation of anisole by acetic anhydride. *Catal. Lett.* 95: 13–17.

59. Rohan, D., Canaff, C., Fromentin, E., and Guisnet, M. 1998. Acetylation of anisole by acetic anhydride over a HBEA zeolite—origin of deactivation of the catalyst. *J. Catal.* 177: 296–305.

60. Guidotti, M., Canaff, C., Coustard, J.-M., Magnoux, P., and Guisnet, M. 2005. Acetylation of aromatic compounds over H-BEA zeolite: the influence of the substituents on the reactivity and on the catalyst stability. *J. Catal.* 230: 375–383.
61. Freese, U., Heinrich, F., and Roessner, F. 1999. Acylation of aromatic compounds on H-Beta zeolites. *Catal. Today* 49: 237–244.
62. Smith, K., Zhenhua, Z., and Hodgson, P. K. G. 1998. Synthesis of aromatic ketones by acylation of aryl ethers with carboxylic anhydrides in the presence of zeolite H-β (H-BEA) in the absence of solvent. *J. Mol. Catal. A: Chem.* 134: 121–128.
63. Beers, A. E. W., van Bokhoven, J. A., de Lathouder, K. M., Kapteijn, F., and Moulijn, J. A. 2003. Optimization of zeolite Beta by steaming and acid leaching for the acylation of anisole with octanoic acid: a structure–activity relation. *J. Catal.* 218: 239–248.
64. Beers, A. E. W., Nijhuis, T. A., Kapteijn, F., and Moulijn, J. A. 2001. Zeolite coated structures for the acylation of aromatics. *Mesoporous Microporous Mater.* 48: 279–284.
65. Lempers, H. E. B. and Sheldon, R. A. 1998. The stability of chromium in CrAPO-5, CrAPO-11, and CrS-1 during liquid phase oxidations. *J. Catal.* 175: 62–69.
66. Spagnol, M., Gilbert, L., Guillot, H., and Tirel, P.-J. 1997. Acylation method for an aromatic compound. WO Patent 9,748,665.
67. Spagnol, M., Gilbert, L., Benazzi, E., and Marcilly, C. 1996. Aromatic ether acylation process. WO Patent 9,635,656.
68. Kantam, M. L., Ranganath, K. V. S., Sateesh, M., Kumar, K. B. S., and Choudary, B. M. 2005. Friedel–Crafts acylation of aromatics and heteroaromatics by Beta zeolite. *J. Mol. Catal. A: Chem.* 225: 15–20.
69. Jaimol, T., Moreau, P., Finiels, A., Ramaswamy, A. V., and Singh, A. P. 2001. Selective propionylation of veratrole to 3,4-dimethoxypropiophenone using zeolite H-Beta catalysts. *Appl. Catal. A: Gen.* 214: 1–10.
70. Gilbert, L. and Spagnol, M. 1997. Aromatic thioether acylation method. WO Patent 9,717,324.
71. Sawant, D. P. and Halligudi, S. B. 2004. Liquid phase acetylation of thioanisole with acetic anhydride to 4-(methylthio) acetophenone (4-MTAP) using H-Beta catalyst. *Catal. Commun.* 5: 659–663.
72. Fromentin, E., Coustard, J.-M., and Guisnet, M. 2000. Mechanism of 1-acetyl-2-methoxynaphthalene isomerisation over an HBea zeolite. *J. Catal.* 190: 433–438.
73. Berreghis, A., Ayrault, P., Fromentin, E., and Guisnet, M. 2000. Acetylation of 2-methoxynaphthalene with acetic anhydride over a series of dealuminated HBea zeolites. *Catal. Lett.* 68: 121–127.
74. Fromentin, E., Coustard, J.-M., and Guisnet, M. 2000. Acetylation of 2-methoxynaphthalene with acetic anhydride over an HBEA zeolite. *J. Mol. Catal. A: Chem.* 159: 377–388.
75. Casagrande, M., Storaro, L., Lenarda, M., and Ganzerla, R. 2000. Highly selective Friedel–Crafts acylation of 2-methoxynaphthlene catalyzed by H-Bea zeolite. *Appl. Catal. A: Gen.* 201: 263–270.
76. Heinichen, H. K. and Hölderich, W. F. 1999. Acylation of 2-methoxynaphthalene in the presence of modified zeolite HBEA. *J. Catal.* 185: 408–414.

77. Harvey, G., Binder, G., and Prins, R. 1995. The contribution of the external surface to the catalytic activity of zeolite Beta. *Stud. Surf. Sci. Catal.* 94: 397–404.
78. Camblor, M. A. and Pérez-Pariente, J. 1991. Crystallization of zeolite Beta: Effect of Na and K ions. *Zeolites* 11: 202–210.
79. Jansen, J. C., Creyghton, E. J., Njo, S. L., van Koningsveld, H., and van Bekkum, H. 1997. On the remarkable behaviour of zeolite Beta in acid catalysis. *Catal. Today* 38: 205–212.
80. Kiricsi, I., Flego, C., Pazzuconi, G., Parker Jr., W. O., Millini, R., Perego, C., and Bellussi, G. 1994. Progress toward understanding Zeolite β acidity: an IR and ^{27}Al NMR spectroscopic study. *J. Phys. Chem.* 98: 4627–4634.
81. Das, D. and Cheng, S. 2000. Friedel–Crafts acylation of 2-methoxynaphthalene over zeolite catalysts. *Appl. Catal. A: Gen.* 201: 159–168.
82. Kantam, M. L., Sateesh, M., Choudary, B. M., Karuri, B. S. R., and Raghavan, K. V. 2001. Improved acylation process for naphthyl ether. JP Patent 2,001,278,832.
83. Moreau, P., Finiels, A., Meric, P., and Fajula, F. 2003. Acetylation of 2-methoxynaphthalene in the presence of Beta zeolites: influence of reaction conditions and textural properties of the catalysts. *Catal. Lett.* 85: 199–203.
84. Andy, P., Garcia-Martinez, J., Lee, G., Gonzalez, H., Jones, C. W., and Davis, M. E. 2000. Acylation of 2-methoxynaphthalene and isobutylbenzene over zeolite Beta. *J. Catal.* 192: 215–223.
85. Kim, S. D., Lee, K. H., Lee, J. S., Kim, J. G., and Yoon, K. E. 2000. The regioselective acylation of 2-methoxynaphthalene to 2-acetyl-6-methoxynaphthalene over zeolite Beta. *J. Mol. Catal. A: Chem.* 152: 33–45.
86. Meric, P., Finiels, A., and Moreau, P. 2002. Kinetics of 2-methoxynaphthalene acetylation with acetic anhydride over dealuminated HY zeolites. *J. Mol. Catal. A: Chem.* 189: 251–262.
87. Harvey, G. and Mäder, G. 1992. The shape-selective acylation of 2-methoxynaphthalene, catalyzed by zeolites Y, Beta and ZSM-12. *Collect. Czech. Chem. Commun.* 57: 862–868.
88. Jaimol, T., Pandey, A. K., and Singh, A. P. 2001. Selective acetylation of toluene to 4-methylacetophenone over zeolite catalysts. *J. Mol. Catal. A: Chem.* 170: 117–126.
89. Sreekumar, R. and Padmakumar, R. 1997. Friedel–Crafts acylation of aromatic hydrocarbons using zeolites. *Synth. Commun.* 27: 777–780.
90. Botta, A., Buysch, H.-J., Puppe, L., and Arlt, D. 1989. Selective etherification of phenyl ketone cpds in 4-position. DE Patent 3,809,260.
91. Botta, A., Buysch, H.-J., Puppe, L., and Arlt, D. 1989. Process for the preparation of phenyl ketones etherified in position 4. EP Patent 0,334,096.
92. Ram Reddy, P., Subrahmanyam, M., and Kulkarni, S. J. 1998. Vapour phase acylation of furan and pyrrole over zeolites. *Catal. Lett.* 54: 95–100.
93. Gauthier, C., Chiche, B., Finiels, A., and Geneste, P. 1989. Influence of acidity in Friedel–Crafts acylation catalyzed by zeolites. *J. Mol. Catal.* 50: 219–229.
94. Ram Reddy, P., Subrahmanyam, M., and Durga Kumari, V. 1999. Vapor-phase synthesis of acetophenone in benzene acylation over CeHZSM-5(30) zeolite. *Catal. Lett.* 61: 207–211.
95. Singh, A. P. and Pandey, A. K. 1997. Acetylation of benzene to acetophenone over zeolite catalysts. *J. Mol. Catal. A: Chem.* 123: 141–147.

96. Pandey, A. K. and Singh, A. P. 1997. A novel catalytic method for the acylation of aromatics to the corresponding ketones over zeolite catalysts. *Catal. Lett.* 44: 129–133.

97. Wang, Q. L., Ma, Y., Ji, X., Yan, H., and Qiu, Q. 1995. Regioselective acylation of anisole with carboxylic acids over HZSM-5 catalyst. *J. Chem. Soc., Chem. Commun.* 2307–2308.

98. Xu, M., Cheng, M., and Bao, X. 2000. Growth of ultrafine zeolite Y crystals on metakaolin microspheres. *Chem. Commun.* 1873–1874.

99. García-Martínez, J., Cazorla-Amorós, D., Linares-Solano, A., and Lin, Y. S. 2001. Synthesis and characterisation of MFI-type zeolites supported on carbon materials. *Microporous Mesoporous Mater.* 42: 255–268.

100. Nijhuis, T. A., Beers, A. E. W., Vergunst, T., Hoek, I., Kapteijn, F., and Moulijn, J. A. 2001. Preparation of monolithic catalysts. *Catal. Rev.* 43: 345–380.

101. Beers, A. E. W., Nijhuis, T. A., Aalders, N., Kapteijn, F., and Moulijn, J. A. 2003. BEA coating of structured supports—performance in acylation. *Appl. Catal. A: Gen.* 243: 237–250.

102. Winé, G., Tessonnier, J. P., Pham-Huu, C., and Ledoux, M. J. 2002. Beta zeolite supported on a macroscopic pre-shaped SiC as a high performance catalyst for liquid-phase benzoylation. *Chem. Commun.* 2418–2419.

103. Ledoux, M. J. and Pham-Huu, C. 2001. Silicon carbide: a novel catalyst support for heterogeneous catalysis. *Cattech* 5: 226–246.

104. Nhut, J.-M., Pesant, L., Tessonnier, J.-P., Winé, G., Guille, J., Pham-Huua, C., and Ledoux, M.-J. 2003. Mesoporous carbon nanotubes for use as support in catalysis and as nanosized reactors for one-dimensional inorganic material synthesis. *Appl. Catal. A: Gen.* 254: 345–363.

105. Choudhary, V. R., Jana, S. K., and Patil, N. S. 2001. Acylation of benzene over clay and mesoporous Si-MCM-41 supported $InCl_3$, $GaCl_3$ and $ZnCl_2$ catalysts. *Catal. Lett.* 76: 235–239.

106. Choudhary, V. R., Jana, S. K., and Patil, N. S. 2002. Acylation of aromatic compounds using moisture insensitive $InCl_3$ impregnated mesoporous Si-MCM-41 catalyst. *Tetrahedron Lett.* 43: 1105–1107.

107. Choudary, V. R., Jana, S. K., Patil, N. S., and Bhargava, S. K. 2003. Friedel–Crafts type benzylation and benzoylation of aromatic compounds over Hβ zeolite modified by oxides or chlorides of gallium and indium. *Microporous Mesoporous Mater.* 57: 21–35.

108. Choudhary, V. R. and Jana, S. K. 2002. Benzoylation of benzene and substituted benzenes by benzoyl chloride over In_2O_3/Si-MCM-41 catalyst. *J. Mol. Catal. A: Chem.* 184: 247–255.

109. Choudhary, V. R., Jana, S. K., and Kiran, B. P. 2000. Highly active Si-MCM-41-supported Ga_2O_3 and In_2O_3 catalysts for Friedel–Crafts-type benzylation and acylation reactions in the presence or absence of moisture. *J. Catal.* 192: 257–261.

110. Reichardt, C. 1988. *Solvents and solvent effects in organic chemistry.* Weinheim, Germany: VCH.

111. Choudhary, V. R. and Mantri, K. 2002. $AlCl_x$-grafted Si-MCM-41 prepared by reacting anhydrous $AlCl_3$ with terminal Si–OH groups: an active solid catalyst for benzylation and acylation reactions. *Microporous Mesoporous Mater.* 56: 317–320.

112. Shih, P.-C., Wang, J.-H., and Moub, C.-Y. 2004. Strongly acidic mesoporous aluminosilicates prepared from zeolite seeds: acylation of anisole with octanoyl chloride. *Catal. Today* 93–95: 365–370.
113. Beers, A. E. W., Nijhuis, T. A., Aalders, N., Kapteijn, F., and Moulijn, J. A. 2003. BEA coating of structured supports—performance in acylation. *Appl. Catal. A: Gen.* 243: 237–250.
114. Gunnewegh, E.A., Gopie, S. S., and van Bekkum, H. 1996. MCM-41 type molecular sieves as catalysts for the Friedel–Crafts acylation of 2-methoxy-naphthalene. *J. Mol. Catal. A: Chem.* 106: 151–158.
115. Balogh, M. and Laszlo, P. 1993. *Organic chemistry using clays*. New York: Springer Verlag.
116. Varma, R. S. 2002. Clay and clay-supported reagents in organic synthesis. *Tetrahedron* 52: 1235–1255.
117. Laszlo, P. 1987. Chemical reactions on clays. *Science* 235: 1473–1477.
118. Brindley, G. W. and Brown, G. 1980. *Crystal structure of clay minerals and their X-ray identification*. London: Mineralogical Society Monograph n. 5, Longman Scientific and Technological.
119. Pinnavaia, T. J. 1983. Intercalated clay catalysts. *Science* 220: 365–371.
120. Thomas, J. M. 1982. In *Intercalation chemistry*, ed. A. J. Jacobsen, 55. London: Academic Press.
121. Corma, A. 1997. From microporous to mesoporous molecular sieve materials and their use in catalysis. *Chem. Rev.* 97: 2373–2420.
122. Cavani, F., Trifirò, F., and Vaccari, A. 1991. Hydrotalcite-type anionic clays: preparation, properties and applications. *Catal. Today* 11: 173–301.
123. Sels, B. F., De Vos. D. E., and Jacobs, P. A. 2001. Hydrotalcite-like anionic clays in catalytic organic reactions. *Catal. Rev. Sci. Eng.* 43: 443–488.
124. Devic, M. and Shirmann, J.-P. 1991. Cyclodéshydratation de l'acide 2-benzoyl-benzoïque en anthraquinone, en milieu sec, sur support solide mineral. *New J. Chem.* 15: 949–953.
125. Bram, G., Loupy, A. Majdoub, M., and Petit, A. 1991. Anthraquinone micro-wave-induced synthesis in dry media in domestic ovens. *Chem. Ind. (London)* 396–397.
126. Acosta, A., de la Cruz, P., De Miguel, P., Diez-Barra, E., de la Hoz, A., Langa, F., Loupy, A., Majdoub, M., Martin, N., Sanchez, C., and Seoane, C. 1995. Microwave assisted synthesis of heterocyclic fused quinones in dry media. *Tetrahedron Lett.* 36: 2165–2168.
127. Clark, J. H., Cullen, S. R., Barlow, S. J., and Bastock, T. W. 1994. Environmentally friendly chemistry using supported reagent catalysts: structure–property relationships for clayzic. *J. Chem. Soc., Perkin Trans.* 2 1117–1130.
128. Cornélis, A., Gerstmans, A., Laszlo, P., Mathy, A., and Zieba, I. 1990. Friedel–Crafts acylations with modified clays as catalysts. *Chem. Lett.* 6: 103–109.
129. Laszlo, P. and Montaufier, M.-T. 1991. A co-reactant reverses relative reactivities. *Tetrahedron Lett.* 32: 1561–1564.
130. Cornélis, A., Laszlo, P., and Wang, S. 1993. On the transition state for "Clayzic"-catalyzed Friedel-Crafts reactions upon anisole. *Tetrahedron Lett.* 34: 3849–3852.
131. Cornélis, A., Laszlo, P., and Wang, S.-F. 1993. Side-product inhibition of the catalyst in electrophilic aromatic substitution and Friedel–Crafts reactions. *Catal. Lett.* 17: 63–69.

132. Chiche, B., Finiels, A., Gauthier, C., Geneste, P., Graille, J., and Pioch, D. 1987. Acylation over cation-exchanged montmorillonite. *J. Mol. Catal.* 42: 229–235.

133. Chiche, B. H., Geneste, P., Gauthier, C., Figueras, F., Fajula, F., Finiels, A., Graille, J., and Pioch, D. 1987. Catalysts based on clay of the smectite type, process for obtaining them and their application to the acylation of aromatic hydrocarbons. FR Patent 2,599,275.

134. Terunori, F. and Kazunori, T. 1986. Production of acyl group-substituted aromatic compound. JP Patent 61,152,636.

135. Bandgar, B. P. and Savadarte, V. S. 1999. Envirocat EPIC® as a novel catalyst for acylation of anisole using benzoic acids. *Synth. Commun.* 29: 2587–2590.

136. Choudary, B. M., Sateesh, M., Lakshmi Kantam, M., and Prasad, K. V. R. 1998. Acylation of aromatic ethers with acid anhydrides in the presence of cation-exchanged clays. *Appl. Catal. A: Gen.* 171: 155–160.

137. Baudry-Barbier, D., Dormond, A., and Duriau-Montagne, F. 1999. Catalytic activity of rare-earth-supported catalysts in Friedel–Crafts acylations. *J. Mol. Catal. A: Chem.* 149: 215–224.

138. Choudary, B. M., Sateesh, M., Lakshmi Kantam, M., Ranganath, K. V. S., and Raghavan, K. V. 2001. Selective acetylation of 5-numbered aromatic heterocycle compounds using metal-exchanged clay catalysts. *Catal. Lett.* 76: 231–233.

139. Biró, K., Békássy, S., Ágai, B., and Figueras, F. 2000. Heterogeneous catalysis for the acetylation of benzo crown ethers. *J. Mol. Catal. A: Chem.* 151: 179–184.

140. López-Salinas, E., Garcia-Sanchez, M., Ramon-Garcia, M. L., and Schifter, I. 1996. New gallium-substituted hydrotalcites: $[Mg_{1-x}Ga_x(OH)_2](CO_3)_{x/2} \cdot mH_2O$. *J. Porous Mater.* 3: 169–174.

141. Choudhary, V. R., Jana, S. K., and Mandale, A. B. 2001. Highly active, reusable and moisture insensitive catalyst obtained from basic Ga-Mg-hydrotalcite anionic clay for Friedel–Crafts type benzylation and acylation reactions. *Catal. Lett.* 74: 95–98.

142. Tauster, S. J. 1987. Strong metal-support interactions. *Acc. Chem. Res.* 20: 389–394.

143. Morley, J. O. 1977. Aromatic acylations catalysed by metal oxides. *J. Chem. Soc., Perkin* 2 601–605.

144. Arata, K., Yabe, K., and Toyoshima, I. 1976. Friedel–Crafts reaction in the heterogeneous system—V. Friedel–Crafts benzylation and benzoylation of toluene catalyzed by calcined iron sulfates. *J. Catal.* 44: 385–391.

145. Sarvari, M. H. and Sharghi, H. 2004. Reactions on a solid surface: a simple, economical and efficient Friedel–Crafts acylation reaction over zinc oxide (ZnO) as a new catalyst. *J. Org. Chem.* 69: 6953–6956.

146. Hesse, M., Hoelderich, W., Dockner, T., and Koehler, H. 1989. Process for the preparation of acylated imidazoles. EP Patent 0,332,075.

147. Bhaskaran, S. K. and Venugopal, T. T. 2001. Liquid-phase benzoylation of *o*-xylene to 3,4-dimethylbenzophenone using rare earth oxide catalysts. *React. Kinet. Catal. Lett.* 74: 99–102.

148. Ranu, B. C., Jana, U., and Majee, A. 1999. Selective monoacylation of ferrocene. An eco-friendly procedure on the solid phase of alumina. *Green Chem.* 1: 33–34.

149. Ranu, B. C., Ghosh, K., and Jana, U. 1996. Simple and improved procedure for regioselective acylation of aromatic ethers with carboxylic acids on the solid surface of alumina in the presence of trifluoroacetic anhydride. *J. Org. Chem.* 61: 9546–9547.

150. Choudhary, V. R. and Jana, S. K. 2002. Acylation of aromatic compounds using moisture insensitive mesoporous Si-MCM-41 suported Ga_2O_3 catalyst. *Synth. Commun.* 32: 2843–2848.

151. Choudhary, V. R. and Jana, S. K. 2001. Highly active and low moisture sensitive supported thallium oxide catalysts for Friedel–Crafts-type benzylation and acylation reactions: strong thallium oxide-support interactions. *J. Catal.* 201: 225–235.

152. Haller, G. L. and Resasco, D. E. 1989. Metal-support interaction: group VIII metals and reducible oxides. *Adv. Catal.* 36: 173–235.

153. Choudary, V. R., Mulla, S. A. R., and Uphade, B. S. 1997. Oxidative coupling of methane over supported La_2O_3 and La-promoted MgO catalysts: influence of catalyst-support interactions. *Ind. Eng. Chem. Res.* 36: 2096–2100.

154. Patil, M. L., Jnaneshwara, G. K., Sabde, D. P., Dongare, M. K., Sudalai, A., and Deshpande, V. H. 1997. Regiospecific acylations of aromatics and selective reductions of azobenzenes over hydrated zirconia. *Tetrahedron Lett.*, 38: 2137–2140.

155. Sakthivel, R., Prescott, H., and Kemnitz, E. 2004. WO_3/ZrO_2: a potential catalyst for the acetylation of anisole. *J. Mol. Catal. A: Chem.* 223: 137–142.

156. Devassy, B. M., Halligudi, S. B., Hegde, S. G., Halgeri, A. B., and Lefebvre, F. 2002. 12-Tungstophosphoric acid/zirconia—a highly active stable solid acid—comparison with a tungstated zirconia catalyst. *Chem. Commun.* 1074–1075.

157. Matsuhashi, H., Kato, K., and Arata, K. 1994. Synthesis of solid superacid of borate supported on zirconium oxides. *Stud. Surf. Sci. Catal.* 90: 251–256.

158. Patil, P. T., Malshe, K. M., Kumar, P., Dongare, M. K., and Kemnitz, E. 2002. Benzoylation of anisole over borate zirconia solid acid catalyst. *Catal. Commun.* 3: 411–416.

159. Deutsch, J., Quaschning, V., Kemnitz, E., Auroux, A., Ehwald, H., and Lieske, H. 2000. Catalytic acylation of aromatics with carboxylic anhydrides over sulfated zirconia. *Top. Catal.* 13: 281–285.

160. Borate, H. B., Gaikwad, A. G., Maujan, S. R., Sawargave, S. P., and Kalal, K. M. 2007. One-step preparation of α-chlorostyrenes. *Tetrahedron Lett.* 48: 4869–4872.

161. Nishamol, K., Rehna, K. S., and Sugunan, S. 2004. Liquid-phase Friedel–Crafts benzoylation of benzene using $Cu_xMn_{(1-x)}Fe_2O_4$ spinel catalysts. *React. Kinet. Catal. Lett.* 81: 229–233.

162. Song, X. and Sayari, A. 1996. Sulfated zirconia-based strong solid-acid catalysts: recent progress. *Catal. Rev. Sci. Eng.* 38: 329–412.

163. Hino, M., Kobayashi, S., and Arata, K. 1979. Solid catalyst treated with anion. 2. Reactions of butane and isobutane catalyzed by zirconium oxide treated with sulfate ion. Solid superacid catalyst. *J. Am. Chem. Soc.* 101: 6439–6441.

164. Hino, M. and Arata, K. 1980. Synthesis of solid superacid catalyst with acid strength of $H_0 \le 16.04$. *J. Chem. Soc., Chem. Commun.* 851–852.

165. Misono, M. and Okuhara, T. 1993. Solid superacid catalysts. *Chemtech* 11: 23–29.

166. Kustov, L. M., Kazansky, V. B., Figueras, F., and Tichit, D. 1994. Investigation of the acidic properties of ZrO_2 modified by SO_2^{-4} anions. *J. Catal.* 150: 143–149.

167. Hino, M. and Arata, K. 1985. Acylation of toluene with acetic and benzoic acids catalysed by a solid superacid in a heterogeneous system. *J. Chem. Soc., Chem. Commun.* 112–113.

168. Arata, K., Nakamura, H., and Shouji, M. 2000. Friedel–Crafts acylation of toluene catalyzed by solid superacids. *Appl. Catal. A: Gen.* 197: 213–219.

169. Deutsch, J., Prescott, H. A., Müller, D., Kemnitz, E., and Lieske, H. 2005. Acylation of naphthalenes and anthracene on sulfated zirconia. *J. Catal.* 231: 269–278.

170. Goto, S., Goto, M., and Kimura, Y. 1990. Benzoylation of toluene by benzoic anhydride on solid superacid catalysts. *React. Kinet. Catal. Lett.* 41: 27–32.

171. Deutsch, J., Trunschke, A., Müller, D., Quaschning, V., Kemnitz, E., and Lieske, H. 2003. Different acylating agents in the synthesis of aromatic ketones on sulfated zirconia. *Catal. Lett.* 88: 9–15.

172. Deutsch, J., Trunschke, A., Müller, D., Quaschning, V., Kemnitz, E., and Lieske, H. 2004. Acetylation and benzoylation of various aromatics on sulfated zirconia. *J. Mol. Catal. A: Chem.* 207: 51–57.

173. Yadav, G. D. and Pujari, A. A. 1999. Friedel–Crafts acylation using sulfated zirconia as a catalyst. Acylation of benzene with 4-chlorobenzoyl chloride over sulfated zirconia as catalyst. *Green Chem.* 1: 69–74.

174. Trunschke, A., Deutsch, J., Müller, D., Lieske, H., Quaschning, V., and Kemnitz, E. 2002. Nature of surface deposits on sulphated zirconia used as catalyst in the benzoylation of anisole. *Catal. Lett.* 83: 271–279.

175. Quaschning, V., Deutsch, J., Druska, P., Niclas, H.-J., and Kemnitz, E. 1998. Properties of modified zirconia used as Friedel–Crafts acylation catalysts. *J. Catal.* 177: 164–174.

176. Hsu, C.-Y., Heimbuch, C. R., Armes, C. Y., and Gates, B. C. 1992. A highly active solid superacid catalyst for *n*-butane isomerization: a sulfated oxide containing iron, manganese and zirconium. *J. Chem. Soc., Chem. Commun.* 1645–1646.

177. Parida, K., Quaschning, V., Lieske, E., and Kemnitz, E. 2001. Freeze-dried promoted and unpromoted sulfated zirconia and their catalytic potential. *J. Mater. Chem.* 11: 1903–1911.

178. Suja, H., Deepa, C. S., Sreeja Rani, K., and Sugunan, S. 2002. Liquid phase benzoylation of arenes over iron promoted sulphated zirconia. *Appl. Catal. A: Gen.* 230: 233–243.

179. Hua, W., Xia, Y., Yue, Y., and Gao, Z. 2000. Promoting effect of Al on SO_4^{2-}/M_xO_y (M = Zr, Ti, Fe) catalysts. *J. Catal.* 196: 104–114.

180. Xia, Y., Hua, W., and Gao, Z. 1998. Benzoylation of toluene with benzoyl chloride on Al-promoted sulfated solid superacids. *Catal. Lett.* 55: 101–104.

181. Sanderson, R. T. 1976. *Chemical bonds and bond energy.* New York: Academic Press.

182. Arata, K. and Hino, M. 1990. Solid catalyst treated with anion—XVIII. Benzoylation of toluene with benzoyl chloride and benzoic anhydride catalysed by solid superacid of sulphate-supported alumina. *Appl. Catal.* 59: 197–204.

183. Jin, T.-S., Yang, M.-N., Feng, G.-L., and Li, T.-S. 2004. Synthesis of diaryl ketones catalyzed by Al_2O_3-ZrO_2/$S_2O_8^{2-}$ solid superacid. *Synth. Commun.* 34: 479–485.

184. Breda, A., Signoretto, M., Ghedini, E., Pinna, F., and Cruciani, G. 2006. Acylation of veratrole over promoted SZ/MCM-41 catalysts: influence of metal promotion. *Appl. Catal. A: Gen.* 308: 216–222.

185. Kozhevnikov, I. V. 1998. Catalysis by heteropoly acids and multicomponent polyoxometalates in liquid-phase reactions. *Chem. Rev.* 98: 171–198.

186. Kozhevnikov, I. V. 2003. Friedel–Crafts acylation and related reactions catalysed by heteropoly acids. *Appl. Catal. A: Gen.* 256: 3–18.

187. Tesser, R., Di Serio, M., Ambrosio, M., and Santacesaria, E. 2002. Heterogeneous catalysts for the production of anthraquinone from 2-benzoylbenzoic acid. *Chem. Eng. J.* 90: 195–201.

188. Kozhevnikov, I. V., Kloetstra, K. R., Sinnema, A., Zandbergen, H. W., and van Bekkum, H. 1996. Study of catalysts comprising heteropoly acid $H_3PW_{12}O_{40}$ supported on MCM-41 molecular sieve and amorphous silica *J. Mol. Catal. A: Chem.* 114: 287–298.

189. Blasco, T., Corma, A., Martínez, A., and Martínez-Escolano, P. 1998. Supported heteropolyacid (HPW) catalysts for the continuous alkylation of isobutane with 2-butene: the benefit of using MCM-41 with larger pore diameters. *J. Catal.* 177: 306–313.

190. Kapustin, G. I., Brueva, T. R., Klyachko, A. L., Timofeeva, M. N., Kulikov, S. M., and Kozhevnikov, I. V. 1990. Acidity of heteropoly acids. *Kinet. Catal.* 31: 1017–1020.

191. Sawant, D. P., Devassy, B. M., and Halligudi, S. B. 2004. Friedel–Crafts benzoylation of diphenyl oxide over zirconia supported 12-tungstophosphoric acid. *J. Mol. Catal. A: Chem.* 217: 211–217.

192. Kaur, J., Griffin, K., Harrison, B., and Kozhevnikov, I. V. 2002. Friedel–Crafts acylation catalysed by heteropoly acids. *J. Catal.* 208: 448–455.

193. Kaur, J., Kozhevnikova, E. F., Griffin, K., Harrison, B., and Kozhevnikov, I. V. 2003. Friedel–Crafts acylation and related reactions catalyzed by heteropoly acids. *Kinet. Catal.* 44: 175–182.

194. Cardoso, L. A. M., Alves, W., Jr., Gonzaga, A. R. E., Aguiar, L. M. G., and Andrade, H. M. C. 2004. Friedel–Crafts acylation of anisole with acetic anhydride over silica-supported heteropolyphosphotungstic acid (HPW/SiO_2). *J. Mol. Catal. A: Chem.* 209: 189–197.

195. Bachiller-Baeza, B. and Anderson, J. A. 2004. FTIR and reaction studies of the acylation of anisole with acetic anhydride over supported HPA catalysts. *J. Catal.* 228: 225–233.

196. López-Salinas, E., Hernández-Cortéz, J. G., Schifter, I., Torres-García, E., Navarrete, J., Gutiérrez-Carrillo, A., López, T., Lottici, P. P., and Bersani, D. 2000. Thermal stability of 12-tungstophosphoric acid supported on zirconia. *Appl. Catal. A: Gen.* 193: 215–225.

197. De Castro, C., Primo, J., and Corma, A. 1998. Heteropolyacids and large-pore zeolites as catalysts in acylation reactions using α,β-unsaturated organic acids as acylating agents. *J. Mol. Catal. A: Chem.* 134: 215–222.

198. Castro, C., Corma, A., and Primo, J. 2002. On the acylation reactions of anisole using α,β-unsaturated organic acids as acylating agents and solid acids as catalysts: a mechanistic overview. *J. Mol. Catal. A: Chem.* 177: 273–280.

199. Izumi, Y., Ogawa, M., and Urabe, K. 1995. Alkali metal salts and ammonium salts of Keggin-type heteropolyacids as solid acid catalysts for liquid-phase Friedel-Crafts reactions. *Appl. Catal. A: Gen.* 132: 127–140.
200. Izumi, Y., Matsuo, K., and Urabe, K. 1983. Efficient homogeneous acid catalysis of heteropoly acid and its characterization through ether cleavage reactions. *J. Mol. Catal.* 18: 299–314.
201. Tagawa, T., Amemiya, J., and Goto, S. 2004. Chlorine-free Friedel–Crafts acylation of benzene with benzoic anhydride on insoluble heteropoly acid catalyst. *Appl. Catal. A: Gen.* 257: 19–23.
202. Kaur, J. and Kozhevnikov, I. V. 2002. Efficient acylation of toluene and anisole with aliphatic carboxylic acids catalysed by heteropoly salt $Cs_{2.5}H_{0.5}PW_{12}O_{40}$. *Chem. Commun.* 2508–2509.
203. Firouzabadi, H., Iranpoor, N., and Nowrouzi, F. 2003. Solvent-free Friedel–Crafts acylation of aromatic compounds with carboxylic acids in the presence of trifluoroacetic anhydride and aluminum dodecatungstophosphate. *Tetrahedron Lett.* 44: 5343–5345.
204. Yadav, G. D., Asthana, N. S., and Kamble, V. S. 2003. Cesium-substituted dodecatungstophosphoric acid on K-10 clay for benzoylation of anisole with benzoyl chloride. *J. Catal.* 217: 88–99.
205. Olah, G. A., Prakash, G. K. S., and Sommer, J. 1985. *Superacids.* New York: Wiley.
206. Tanabe, K., Hattori, H., and Yamaguchi, T. 1990. Surface properties of solid superacids. *Crit. Rev. Surf. Chem.* 1: 1–25.
207. Olah, G. A., Malhotra, R., Narang, S. C., and Olah, J. A. 1978. Heterogeneous catalysis by solid superacids. 11. Perfluorinated resinsulfonic acid (Nafion-H) catalyzed Friedel–Crafts acylation of benzene and substituted benzenes. *Synthesis* 672–673.
208. Olah, G. A., Mathew, T., Farnia, M., and Prakash, G. K. S. 1999. Nafion-H catalysed intramolecular Friedel–Crafts acylation: formation of cyclic ketones and related heterocycles. *Synlett* 1067–1068.
209. Yamato, T., Hideshima, C., Prakash, G. K. S., and Olah, G. A. 1991. Organic reactions catalyzed by solid superacids. 5. Perfluorinated sulfonic acid resin (Nafion-H) catalyzed intramolecular Friedel–Crafts acylation. *J. Org. Chem.* 56: 3955–3957.
210. Heidekum, A., Harmer, M. A., and Hoelderich, W. F. 1999. Nafion/silica composite material reveals high catalytic potential in acylation reactions. *J. Catal.* 188: 230–232.
211. Beers, A. E. W., Hoek, I., Nijhuis, T. A., Downing, R. S., Kapteijn, F., and Moulijn, J. A. 2000. Structured catalysts for the acylation of aromatics. *Top. Catal.* 13: 275–280.
212. Alvaro, M., Corma, A., Das, D., Fornés, V., and García, H. 2005. "Nafion"-functionalized mesoporous MCM-41 silica shows high activity and selectivity for carboxylic acid esterification and Friedel–Crafts acylation reactions. *J. Catal.* 231: 48–55.
213. Sarsani, V. R., Lyon, C. J., Hutchenson, K. W., Harmer, M. A., and Subramaniam, B. 2007. Continuous acylation of anisole by acetic anhydride in mesoporous solid acid catalysts: reaction media effects on catalyst deactivation. *J. Catal.* 245: 184–190.
214. Kodomari, M., Suzuki, Y., and Yoshida, K. 1997. Graphite as an effective catalyst for Friedel–Crafts acylation. *Chem. Commun.* 1567–1568.

215. Sarvari, M. H. and Sharghi, H. 2005. Solvent-free catalytic Friedel–Crafts acylation of aromatic compounds with carboxylic acids by using a novel heterogeneous catalyst system: *p*-toluenesulfonic acid/graphite. *Helv. Chim. Acta* 88: 2282–2287.
216. Pârvulescu, A. N., Gagea, B. C., Pârvulescu, V. I., De Vos, D., and Jacobs, P. A. 2006. Acylation of 2-methoxynaphthalene with acetic anhydride over silica-embedded triflate catalysts. *Appl. Catal. A: Gen.* 306: 159–164.
217. Pârvulescu, A. N., Gagea, B. C., Pârvulescu, V., Pârvulescu, V. I., Poncelet, G., and Grange, P. 2002. Comparative behavior of silica-embedded *tert*-butyldimethylsilyltrifluoro-methanesulfonate and lanthanum triflate catalysts. *Catal. Today* 73: 177–185.
218. Harmer, M. A., Sun, Q., Michalczyk, M. J., and Yang, Z. 1997. Unique silane modified perfluorosulfonic acids as versatile reagents for new solid acid catalysts. *Chem. Commun.* 1803–1804.
219. Harmer, M. A., Junk, C., Rostovtsev, V., Carcani, L. G., Vickery, J., and Schnepp, Z. 2007. Synthesis and applications of superacids. 1,1,2,2-Tetrafluoroethanesulfonic acid, supported on silica. *Green Chem.* 9: 30–37.
220. Mitall. K. L. 1992. *Silane and Other Coupling Agents*. Netherlands: VSP.
221. Hu, R.-J. and Li, B.-G. 2004. Novel solid acid catalyst, bentonite-supported polytrifluoromethanesulfosiloxane for Friedel–Crafts acylation of ferrocene. *Catal. Lett.* 98: 43–47.
222. Chidambaram, M., Curulla-Ferre, D., Singh, A. P., and Anderson, B. G. 2003. Synthesis and characterization of triflic acid-functionalized mesoporous Zr-TMS catalysts: heterogenization of CF_3SO_3H over Zr-TMS and its catalytic activity. *J. Catal.* 220: 442–456.
223. Chidambaram, M., Venkatesan, C., Rajamohanan, P. R., and Singh, A. P. 2003. Synthesis of acid functionalized mesoporous $\equiv$Zr---O---SO_2---CF_3 catalysts; heterogenization of CF_3SO_3H over mesoporous $Zr(OH)_4$. *Appl. Catal. A: Gen.* 244: 27–37.
224. Chidambaram, M., Curulla-Ferre, D., Singh, A. P., and Anderson, B. G. 2003. Synthesis and characterization of triflic acid-functionalized mesoporous Zr-TMS catalysts: heterogenization of CF_3SO_3H over Zr-TMS and its catalytic activity. *J. Catal.* 220: 442–456.
225. Landge, S. M., Chidambaram, M., and Singh, A. P. 2004. Benzoylation of toluene with *p*-toluoyl chloride over triflic acid functionalized mesoporous Zr-TMS catalyst. *J. Mol. Catal. A: Chem.* 213: 257–266.
226. Yadav, G. D. and Bhagat, R. D. 2005. Experimental and theoretical analysis of Friedel–Crafts acylation of thioanisole to 4-(methylthio)acetophenone using solid acids. *J. Mol. Catal. A: Chem.* 235: 98–107.
227. Yadav, G. D. and Pimparkar, K. P. 2007. Insight into Friedel–Crafts acylation of 1,4-dimethoxybenzene to 2,5-dimethoxyacetophenone catalysed by solid acids—mechanism, kinetics and remedies for deactivation. *J. Mol. Catal. A: Chem.* 264: 179–191.
228. Melero, J. A., van Grieken, R., Morales, G., and Nuño, V. 2004. Friedel Crafts acylation of aromatic compounds over arenesulfonic containing mesostructured SBA-15 materials. *Catal. Commun.* 5: 131–136.
229. Selvaraj, M., Lee, K., Yoo, K. S., and Lee, T. G. 2005. Synthesis of 2-acetyl-6-methoxynaphthalene using mesoporous SO_4^{2-}/Al-MCM-41 molecular sieves. *Microporous Mesoporous Mater.* 81: 343–355.

chapter 5

Direct phenol acylation
The Fries rearrangement

The direct acylation of phenols represents an extremely complex reaction. Phenol is a typical ambidental system and reacts with acylating reagents in the presence of convenient homogeneous or heterogeneous catalysts to give esters by O-acylation as well as ketones by C-acylation of the aromatic ring. Furthermore, phenyl esters can undergo the Fries rearrangement, complicating the entire process.

The O-acylation process is much more rapid than the C-acylation one and, in general, the *ortho*-hydroxyaryl ketones are prevalent with respect to para-isomers.[1] This is probably due to the ortho-directing effect of the OH group, which can stabilize by hydrogen bond the transition state leading to the ortho attack (Scheme 5.1) in a way resembling the complex-induced proximity effect (CIPE).[2]

The production of aromatic hydroxyketones can also be performed by the Fries rearrangement; in this case, the mode of para-acylation is probably different from that of ortho-acylation. Indeed, the ortho-isomer is a primary product, whereas the para-isomer seems to be a secondary product. Of course, other methods for

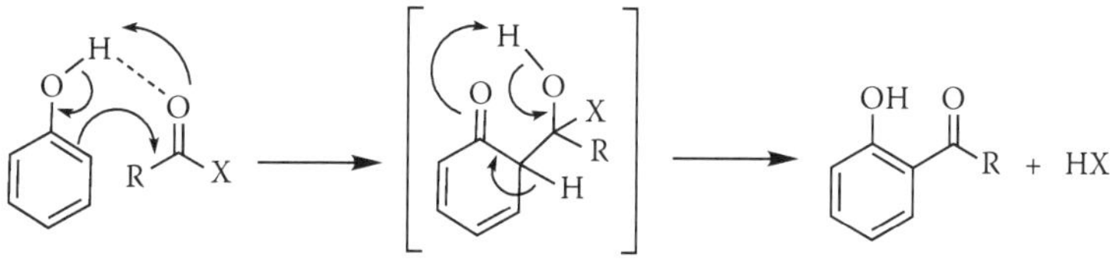

R = alkyl, aryl
X = Cl, OCOR, OH

Scheme 5.1

the formation of *ortho*-hydroxyaryl ketones can result from the acylation of phenols with phenyl esters, which are better acylating agents than carboxylic acids.

Results from the literature suggest that direct phenol acylation and the Fries rearrangement are frequently competitive processes and difficult to characterize by the mechanistic point of view. Consequently, in Section 5.1, we include the synthetic process where the phenol substrate, the acylating agent, and the catalyst are mixed together in the starting reaction mixture aside from the specific reaction mechanism, whereas in Section 5.2 we include reactions involving phenyl esters.

5.1 Direct phenol acylation

5.1.1 Metal phenolates

The use of aluminum triphenolates such as **1** in toluene results in exclusive direct ortho-C-acylation with monochloroacetyl chloride. Indeed, the high coordinating ability of the metal can aid in promoting the formation of an organized reacting complex **2** that undergoes electrophilic attack to the proximate ortho position of the phenol ring, giving product **3** in 71% yield and 98% selectivity (Scheme 5.2).[3]

In the case of bromomagnesium 2-*tert*-butylphenolate and different acyl chlorides, the phenol oxygen competes with the *ortho*-carbon, as previously mentioned, affording a mixture of esters and *ortho*-ketones. However, the electron-withdrawing power of the group at the α-position

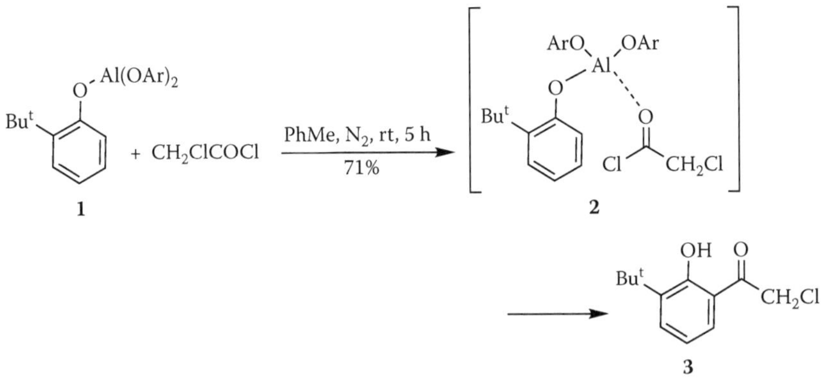

Scheme 5.2

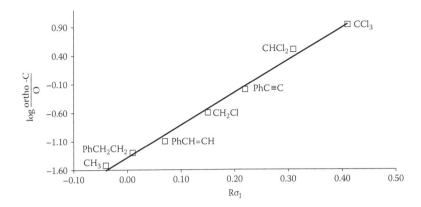

Figure 5.1 Electron-withdrawing power effect of R group in RCOCl on the log ortho-C/O reactivity ratio.

in RCOCl has been shown to play a dramatic role in the ortho-C/O acylation ratio (Figure 5.1).

An increase in the electron-withdrawing power of R in RCOCl results in a gradual increase in the ortho-C/O reactivity ratio. This effect is observed in spite of the obvious increase in size and increase in the steric requirements of the reagents involved (see, for example, the series R = Me, CH_2Cl, $CHCl_2$, CCl_3). There is a reasonable correlation between the logarithm of the ortho-C/O reactivity ratio and the polar substituent constant σ_I, which is considered to be the true measure of the inductive effect of the substituent R group.[4]

From a synthetic point of view, the process can be efficiently applied to the preparation of a large variety of *ortho*-hydroxyaryl ketones, as summarized in Table 5.1.[3,5] Results from Table 5.1 are a proof of the powerful metal-template effect in promoting ortho-regioselective electrophilic acylation of phenol substrates.

A further application of the metal-template effect in the ortho-regioselective acylation of phenols is represented by the direct synthesis of salicyloyl chloride and derivatives by the reaction of bromomagnesium phenolates **4** with phosgene.[6] The reaction affords the unstable salicyloyl chloride–magnesium complexes **5** by a pathway similar to the mechanism depicted in Scheme 5.2. These intermediates can be in situ converted into the corresponding acids **6** (Scheme 5.3), esters, amides, or ketones by reaction with suitable reagents.

5.1.2 Metal halides

The metal-template effect is also exploited as a powerful instrument to achieve highly selective ortho-acylation of the phenol ring in the

Table 5.1 Ortho-regioselective acylation of
phenoxymagnesium bromides with acyl chlorides

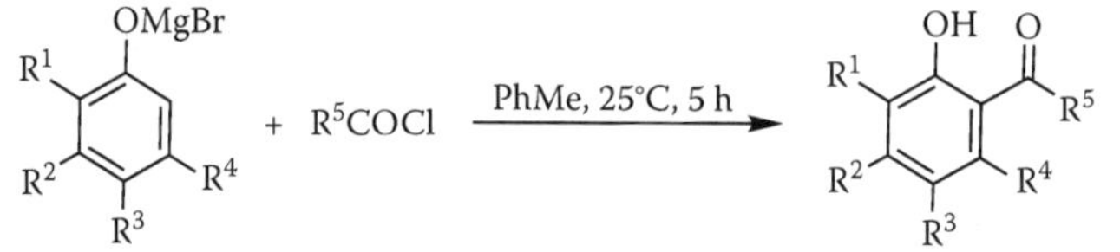

R¹	R²	R³	R⁴	R⁵	Yield (%)
H	OCH₂O		H	PhCH=CH	70
Buᵗ	H	H	H	3,4(OCH₂O)C₆H₃CH=CH	55
H	OCH₂O		H	3,4(OCH₂O)C₆H₃CH=CH	60
H	OMe	H	H	Me(CH=CH)₂	70
H	OCH₂O		H	Me(CH=CH)₂	78
H	H	(CH=CH)₂		Me(CH=CH)₂	58
H	OH	Me	H	Me(CH=CH)₂	75
H	OCH₂O		H	PhC≡C	89

reaction with trichloroacetonitrile as acylating agent according to a typical Houben–Hoesch reaction (Scheme 5.4).[7] The best catalyst is boron chloride, which, by reacting with phenols in 1,2-dichloroethane, produces dichlorophenoxyboron derivatives, the actual reactive species responsible for the activation of trichloroacetonitrile as well as the ortho-regioselective attack.

Two acyl groups can be consecutively introduced on polyhydroxyben-zenes by carefully selected experimental conditions. Thus, 5-acetyl-2,4-di-hydroxyacetophenone is synthesized in 90% yield in a single step from

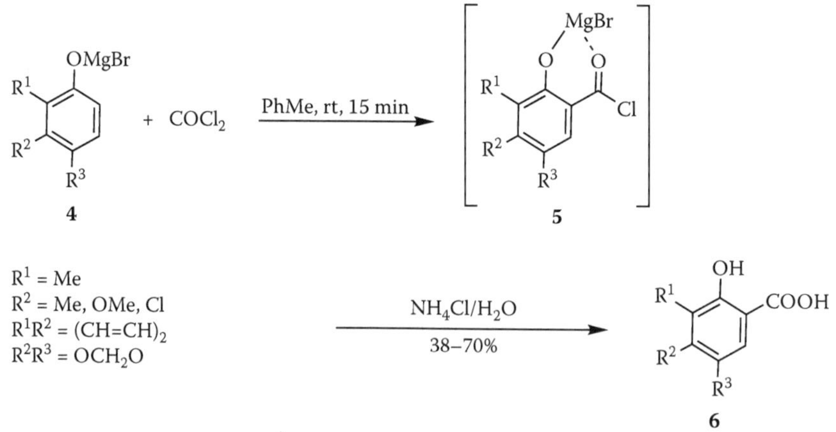

R^1 = Me
R^2 = Me, OMe, Cl
R^1R^2 = (CH=CH)₂
R^2R^3 = OCH₂O

Scheme 5.3

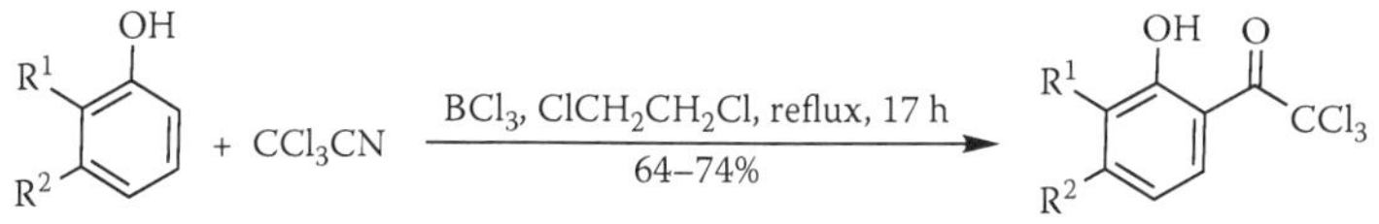

R^1 = H, Me
R^2 = H, Bu^t

Scheme 5.4

resorcinol and zinc chloride–acetic anhydride (AAN) reagent without any solvent. It must be emphasized that lower yield and selectivity are achieved by using iron trichloride–acetyl chloride (AC) mixture under similar conditions.[8]

The positive effect of zinc chloride-AAN reagent in the acylation of phenol derivatives can be exploited in the production of various poly-hydroxyacetophenones (Table 5.2). When the reagent is applied to acyla-tion of phenol, an equimolecular mixture of 2- and 4-hydroxyacetophenone (36% and 40% yield, respectively) is produced. This result allows exclusion of any remarkable metal-template effect in the present system.

Table 5.2 Zinc-chloride-promoted acetylation of polyhydroxybenzenes with AAN at 145°C–150°C for 15 min

Reagent	Reagent/(MeCO)$_2$O/ZnCl$_2$	Product	Yield (%)
	1/2/2	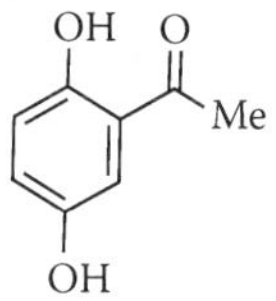	76
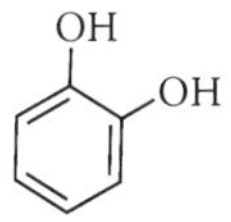	1/3/3	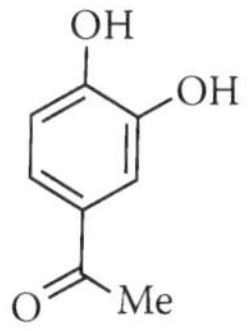	51
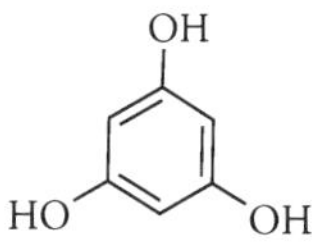	1/3/3	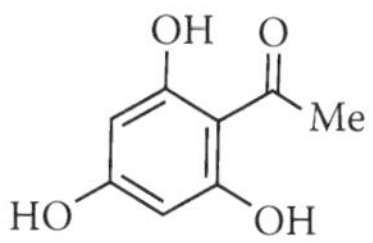	50
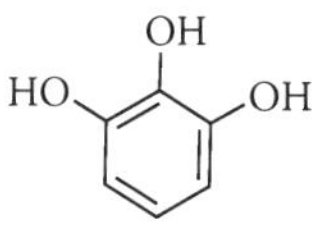	1/3/3	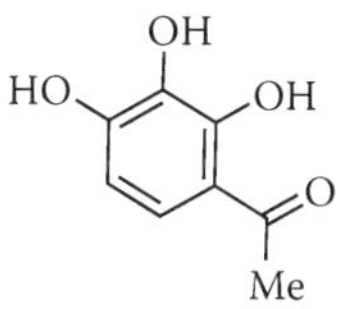	53

Table 5.3 Scandium-triflate-catalyzed acylation of aromatics with acyl chlorides

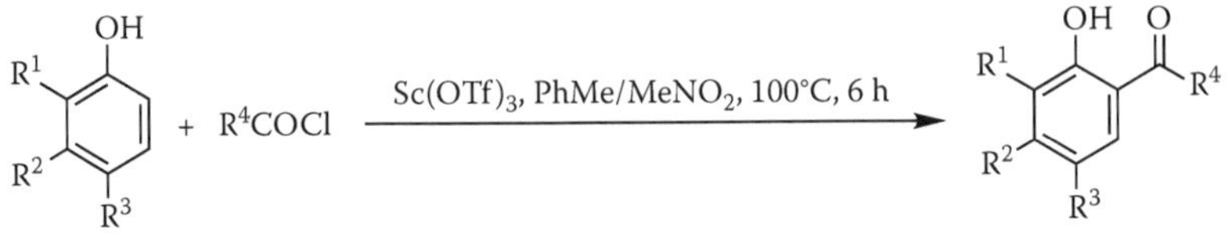

$$R^1\text{-Ar(OH)}R^2R^3 + R^4COCl \xrightarrow{\text{Sc(OTf)}_3,\ \text{PhMe/MeNO}_2,\ 100°C,\ 6\ h} R^1\text{-Ar(OH)}(COR^4)R^2R^3$$

R^1	R^2	R^3	R^4	Yield (%)
H	OMe	H	Me	74
H	Me	H	Me	62
CH=CH-CH=CH		H	Me	93
CH=CH-CH=CH		H	Et	98
CH=CH-CH=CH		H	C_5H_{11}	90
CH=CH-CH=CH		H	c-C_6H_{11}	92
CH=CH-CH=CH		OMe	Me	93
CH=CH-CH=CH(OMe)		H	Me	73
CH=CH-CH=CH(OH)		H	Me	68
CH=CH-CH(OH)=CH		H	Me	78

5.1.3   Metal triflates

Metal triflates are efficiently utilized as reusable catalysts in the direct acylation of phenols as well as in the Fries rearrangement.[9–11] Scandium triflate is employed as catalyst (5% mol) in the direct acylation of phenols and naphthols with acyl chlorides in toluene–nitromethane mixtures at 100°C for 6 h. Results from Table 5.3 show that the reaction is very efficient: complete regioselectivity is observed with *meta*-cresol and 1-naphthol derivatives.

The high catalytic activity of scandium triflate depends on the fact that this compound is not trapped by the free hydroxy groups of the phenols as well as the carbonyl oxygens of the products in a different way from what happens with other classic Lewis acids such as aluminum chloride.

By using hafnium triflate (0.1% mol) in lithium perchlorate/nitromethane mixture at 50°C for 6 h, phenols undergo para-regioselective acetylation with AC in 64% yield, whereas *meta-tert*-butylphenol gives ortho-acylation in 76% yield at the less hindered position.[12]

5.1.4   Zeolites

Studies on phenol acylation with acetic acid (AAC) over ZSM-5 zeolite confirm that the reaction is very complicated.[1,13] The reaction is carried out in a flow reactor charged with the zeolite catalyst at 280°C. The main

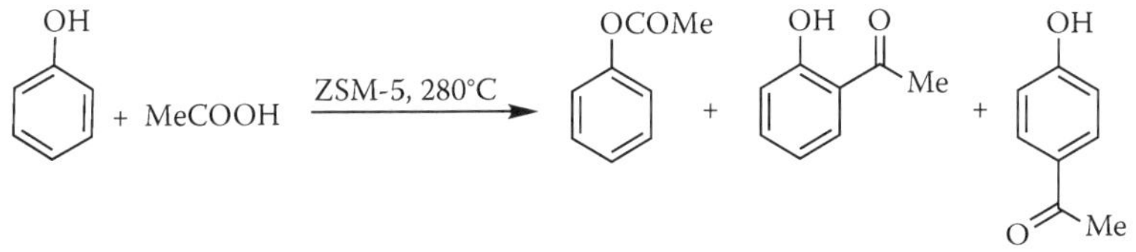

Scheme 5.5

reaction products are phenyl acetate (PA), *ortho*-hydroxyacetophenone (*ortho*-HAP), and *para*-hydroxyacetophenone (*para*-HAP) (Scheme 5.5).

Some acetone is also produced from AAC according to the equation 2 MeCOOH → MeCOMe + CO_2 + H_2O. *ortho*-HAP represents the major isomer (initial yield = 16%) with respect to *para*-HAP (initial yield = 0.8%). By increasing the TOS, an increase in PA yield together with a decrease in *ortho*-HAP yield is observed.

Dealumination of the ZSM-5 zeolite[14] shows a great effect on ortho/ para selectivity in the acylation of phenol by AAC.[13] Thus, for a phenol conversion of 20%, ortho/para selectivity is 7.0 when ZSM-5(41.8) is utilized, and becomes 13.0 in the presence of ZSM-5(42.4). This unexpected increase in the ortho/para ratio can only be explained by postulating that *ortho*-HAP and *para*-HAP result from different pathways. The ortho isomer is mainly produced in the pores, whereas para isomer production occurs only on the external acid sites. The ortho isomer can be formed by direct C-acylation of phenol with AAC; this selective reaction can be related to the general mechanism reported in Scheme 5.1. *ortho*-HAP can also be obtained by the Fries rearrangement. On the contrary, the para-isomer is a secondary product and, therefore, it results from the acylation of phenol by PA according to Scheme 5.6.

In the same reaction, the Y zeolite shows the highest initial selectivity values to *ortho*-HAP (69%),[15] but after 4 h it decreases to only 10%. The most active catalyst still is confirmed to be the ZSM-5 zeolite; in fact, with this catalyst, no significant activity decay is observed (from 67% to 65%) for the *ortho*-HAP formation rate during catalytic runs. ZSM-5 does not deactivate by increasing the conversion of phenol up to 25%, contrary to what is observed for the Y sample. The superior stability of ZSM-5 can be

Scheme 5.6

explained by supposing that its small pore size structure could hamper the formation of coke precursor compounds.

ZSM-5 zeolites modified by inclusion of metal cations show higher activity than the starting materials in the acylation of phenol with AAN.[16] Reactions are carried out using a tubular downflow reactor heated at 250°C. The most promising catalyst is CeZSM-5, showing 75% phenol conversion with a 32.33 ortho/para acylation ratio. An increase in the SAR value results in an increase of O-acylation, suggesting that C-acylation requires more Brönsted acidity.

Unmodified BEA zeolite, on the other hand, shows the highest activity in the acylation of phenol with benzoic anhydride.[17] Phenyl benzoate (PB) is the main product (61% yield), accompanied by C-acylated products (35%), with an interesting para-selectivity (ortho/para = 0.48). When the reaction time is increased from 4 to 20 h, an increase in *para*-hydroxybenzophenone yield (from 11% to 23%) together with a decrease in PB yield (from 79% to 64%) is observed; however, a small increase in the *ortho*-hydroxybenzophenone yield (from 9% to 10%) cannot be avoided. The activity of the catalyst, together with its selectivity, does not distinctly decrease when the catalyst is used from fresh to first recycle.

Phenol acylation with AAC and AAN has been studied in batch conditions over mesoporous AlMCM-41 catalyst.[18] PA is the major product and, in all experiments, *ortho*-HAP is the predominant C-acylated product, without any detectable amount of *para*-HAP; instead, a double acylated product is isolated in trace amounts. Different behaviors are observed upon increasing the temperature with the two acylating agents: when AAN is utilized, the overall conversion of the phenol decreases with temperature increase; this is ascribed to the decomposition of the PA back to phenol, which results in an apparent conversion decline (from 99% to 52%). At the same time, the selectivity for *ortho*-HAP increases from 1% to 21%, indicating that, at lower temperatures, O-acylation is the predominant process, but as the temperature increases, O-acylation is gradually replaced by C-acylation. When AAC is utilized, the overall conversion is much lower than that obtained with AAN, but it increases from 33% to 47%, together with *ortho*-HAP selectivity (from 1% to 24%) by raising the temperature from 200 to 425°C. Finally, an increase in the SAR value results in a lowering of the phenol conversion, whereas the selectivity of *ortho*-HAP and double acetylated products is enhanced.

Phenol undergoes propanoylation with propanoyl chloride over BEA, Y, MOR, and ZSM-5 zeolites.[19] The main product is phenylpropanoate; moreover, BEA (Scheme 5.7) is the most active in the C-acylation and selective for *para*-hydroxypropiophenone production (ortho/para = 0.53) compared to other zeolite catalysts.

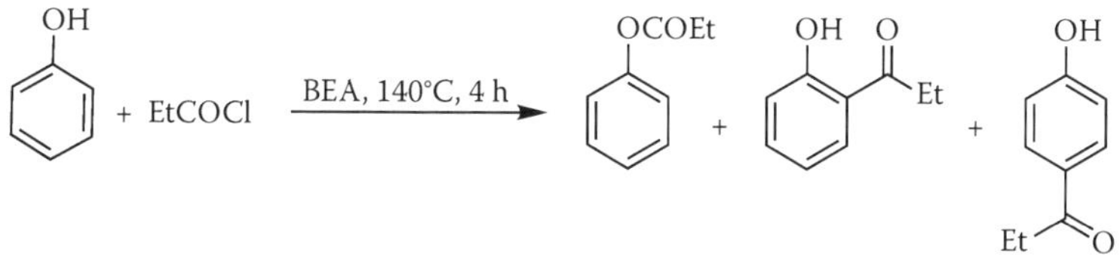

Scheme 5.7

The conventional catalyst aluminum chloride is less active and selective (ortho/para = 1.11), and a higher yield of other by-products (4%) is obtained due to its non-shape-selective character.

The CeY zeolite is utilized for the preparation of 4-methylcoumarin by the reaction of phenol with AAN.[20] The formation of PA represents the first step; the subsequent acylation at the ortho position, followed by an intramolecular aldol-like condensation, affords the final 4-methylcoumarin in 75% yield (Scheme 5.8). In the entire process, the cerium-catalyst shows a bifunctional character: the active centers in the supercage of CeY zeolite, the Ce^{3+} ions, act as Lewis acid catalysts, whereas the acid centers H^+, formed by the dissociation of water according to the equation $Ce^{3+} + H_2O \rightarrow [Ce(OH)]^{2+} + H^+$, act as Brönsted catalysts.

Studies on catalyst deactivation during phenol acylation with PA in the presence of BEA zeolite were performed by recovering the organic material entrapped into the zeolite following two methodologies: (1) extraction by Soxhlet of the zeolite [Ext] and (2) extraction of coke by dissolution of the zeolite itself in a 40% solution of hydrofluoric acid [Coke].[21] The model acylation reaction is carried out in the two classical solvents, dodecane and sulfolane. In both cases, a significant lowering of the rate of formation of HAP with time is observed: this deactivation is faster in dodecane (~1 h) than in sulfolane (~2 h). Whatever the solvent, the two reactants are the main components of the material retained in the catalyst; nevertheless, in the case of sulfolane, their content in Ext and

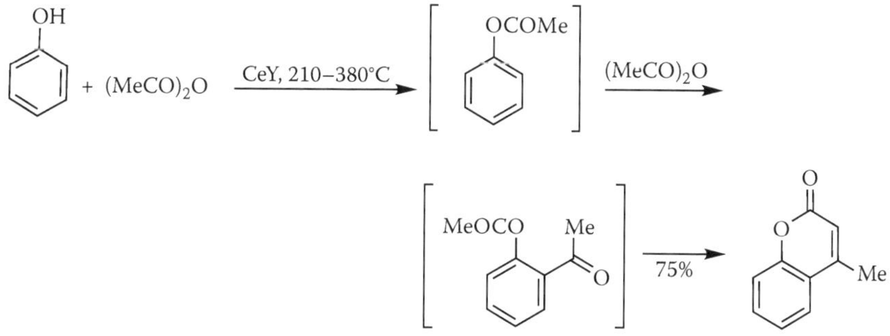

Scheme 5.8

Coke is similar to that of the reaction mixture, whereas when the less polar solvent dodecane was utilized, their content in Ext and Coke was greater than that in the reaction mixture. In addition, sulfolane constitutes a very significant part of Ext and Coke, whereas practically no dodecane is found. As already stressed in this book, these differences can be related to the difference in solvent polarity: the polar sulfolane enters the pores of the zeolite, whereas dodecane cannot enter them. For the same reason, the amount of PA found in Coke is very small compared to the more polar phenol and sulfolane. Catalyst deactivation by product inhibition is also confirmed; the reaction rate is very low when PA is added to the zeolite impregnated with phenol, the higher rate being obtained when phenol is added to the zeolite impregnated with the less polar PA. The conclusion is that the decrease in catalyst activity is not necessarily due to the formation of the heavy secondary products but is most likely due to limited access of PA to the zeolite pores occupied by the very polar phenol.

The benzoylation of resorcinol to produce 2,4-dihydroxybenzo-phenone has been previously performed with benzotrichloride, which implies the coproduction of 3 mol of hydrochloric acid and consequently a large amounts of acid waste.[22–24] Alternative routes to 2,4-dihydroxyben-zophenone involve direct acylation of resorcinol with benzoyl chloride (BC) or the Hoesch reaction with benzonitrile, which also suffer from acid waste production.[25,26] In a more ecoefficient approach, the reaction was performed with benzoic acid (BAC) in the presence of some solid cata-lysts, with a special focus on BEA zeolite (Scheme 5.9).[27,28] The progress of the reaction in *para*-chlorotoluene is examined, and after 18 h, 2,4-dihy-droxybenzophenone (2,4-DHB) is isolated in 70% yield, together with 20% of resorcinol monobenzoate (RMB), 3% of resorcinol dibenzoate, 5% of BAC, and 2% of resorcinol. The fact that the concentration of resorcinol in the final mixture is somewhat lower than that of BAC is probably due to the preferred adsorption of the former. The process is then studied by using different substituted BACs: the different conversions of 2-methyl-,

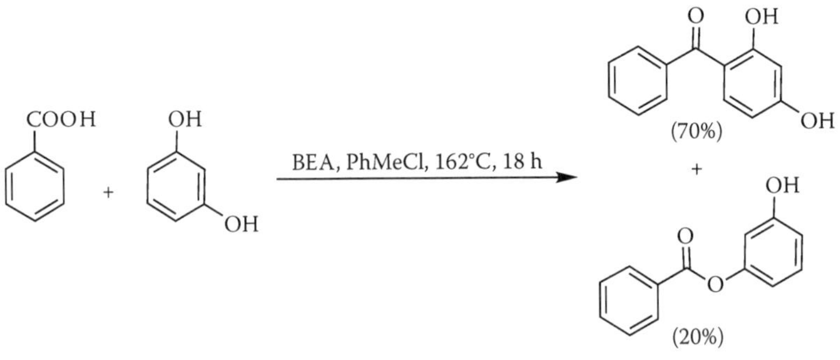

Scheme 5.9

4-methyl- and 2,6-dimethylbenzoic acids (70%, 80%, and 28%, respectively) are ascribed to transition state shape selectivity; this is confirmed by the higher yield (98%) observed when the resorcinol is reacted with 2,6-dimethylbenzoic acid in the presence of Amberlyst-15, a strongly acidic cation-exchanged resin.

Benzotrichloride and BC are, however, efficiently utilized for small-scale hydroxybenzophenone synthesis.[29] Thus benzoylation of phenol can be simply performed by stirring at 120°C for 5 h an equimolecular mixture of phenol and benzotrichloride in the presence of ZSM-5 in 1,2-dichloroethane. 4-Hydroxybenzophenone is obtained in 80% yield, whereas by using BC under the same reaction conditions, the same product is isolated in 67% yield.

5.1.5 Clays

Clays are utilized as catalysts for the direct acylation of resorcinol with phenyl acetyl chloride.[30–32] In all cases, high resorcinol conversion is achieved (65%–80%), and phenyl acetyl chloride reacts completely. However, the yield of the desired ketone is poor, and the best yield (25%) is achieved with KS montmorillonite, ester being the main product. The ketone/ester ratio is practically constant during the reaction's progress with every catalyst. Moreover, the concentration of the ester shows no maximum, which means that the ketone is not formed by a rearrangement of the ester but directly by ring acylation. The FeK10 and FeKS clays are a little more active for C-acylation than the undoped ones (KS: ester yield = 39%, ketone yield = 25%; FeKS: ester yield = 37%, ketone yield = 42%). In the presence of FeK10 without solvent in a melted phase, although resorcinol conversion is a little lower (~70%) than in 1,2-dichloroethane (80%), good ketone selectivity makes possible a 60% preparative yield of product **7** (Scheme 5.10).

The solvent-free reaction has some other advantages: in fact, although the catalysts used in 1,2-dichloroethane are generally deactivated after one cycle, the solvent-free reaction shows a satisfactory catalyst reusability after washing and drying (80% resorcinol conversion in the second cycle).

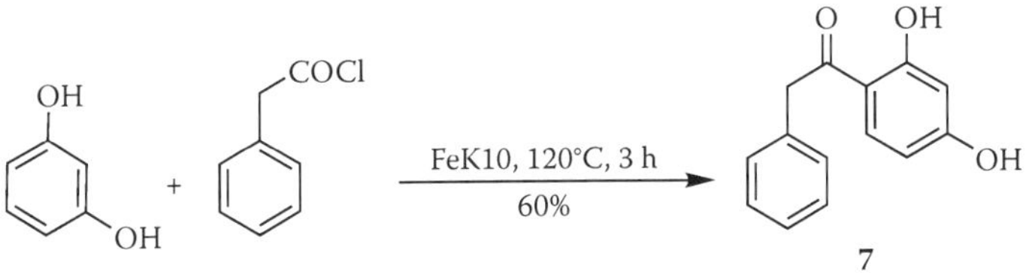

Scheme 5.10

BAC is also utilized as a resorcinol acylating agent in the presence of montmorillonite Fulcat 22B.[33] RMB and 2,4-DHB are the sole products; as already observed for phenol, the former is claimed to be a primary product, whereas the latter is a secondary one, being exclusively formed by the Fries rearrangement. These results seem to be not in complete agreement with other mechanistic conclusions,[30] probably because they are obtained with a different acylating agent. On the basis of these data, it is possible to conclude that RMB first diffuses out of the catalyst pores and is successively activated and transformed into 2,4-DHB. One possible explanation is that the high hydrophilic character of the clay surface makes the pores fill with more polar substances, such as BAC and resorcinol, while less polar compounds rapidly diffuse out of the clay porosity and undergo consecutive transformations at the active sites located at the external particle surface. The reaction between resorcinol and BAC to yield RMB is fully reversible, and hydrolysis of RMB to yield resorcinol and BAC occurs with water retained in clay; the removal of water from the bulk liquid by azeotropic distillation makes it possible to considerably increase conversion, which reaches 81% because the esterification reaction is favored; in this case the 2,4-DHB is obtained with 52% yield (62% selectivity).

5.1.6   Graphite

Graphite combined with methanesulfonic acid represents an effective catalyst for the acylation of phenols and naphthols, with carboxylic acids giving a good yield of *ortho*-hydroxy aryl ketones[34] (Table 5.4). The reactions seem to be faster with activated aromatic carboxylic acids. It must be emphasized that, in the absence of graphite, in the model reaction between *meta*-cresol and BAC, the product is isolated in lower yield (20% versus 81%).

The recovered graphite shows good catalytic activity on recycling: the yields of the ortho-product in the model reaction in the second, third, and fourth cycles are almost the same as the yield in the first run. Because methanesulfonic acid is not adsorbed onto graphite, it is necessary to add it again in the successive cycle after having washed the catalyst with methylene chloride.

5.2   The Fries rearrangement

The Fries reaction consists of the catalyzed rearrangement of phenyl esters with the production of hydroxy aryl ketones[35] (Scheme 5.11). From analysis of a great number of papers dealing with the Fries reaction promoted under both homogeneous and heterogeneous catalysis, it becomes clear that mechanistic problems are more and more entwined. In order to help the reader draw general conclusions from the wealth of sometimes

Table 5.4 Promotion of acylation of phenols with carboxylic acids by graphite–methanesulfonic acid

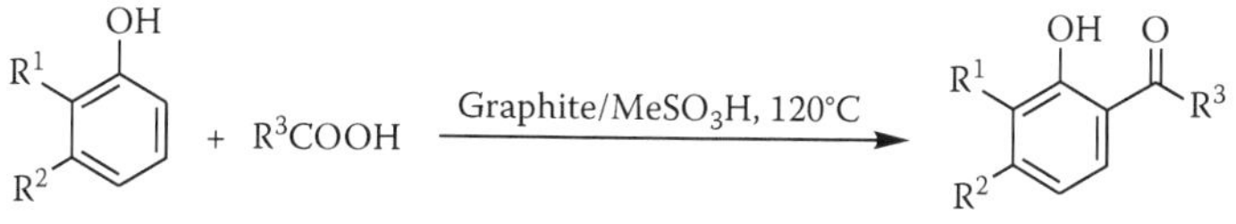

R^1	R^2	R^3	t (h)	Yield (%)
H	Me	Ph	3	81
H	Me	$2\text{-ClC}_6\text{H}_4$	3.5	80
H	Me	$3\text{-MeC}_6\text{H}_4$	2	86
H	Me	$c\text{-C}_6\text{H}_{11}$	2	90
H	Me	Bn	2	71
H	Me	Me	2	80
H	Me	$\text{Br(CH}_2)_{11}$	2	77
H	F	Me	4	60
H	OH	Me	2.5	80
CH=CH-CH=CH		Me	3	61
CH=CH-CH=CH		$\text{Me(CH}_2)_4$	3	81

conflicting information from the chemical literature, we have summarized some general features, allowing a possible parallel to be drawn between the reaction promoted by classic Lewis acids (i.e., aluminum chloride) and by solid catalysts: (1) the yield of *ortho*-HAP is greater at high temperatures than in the reaction at low temperatures, other conditions being equal; (2) the ortho/para ratio increases with time in the reaction carried out with lower amount of Lewis or solid acids, and it remains constant with higher amount of acid catalyst; and (3) solvents of great polarity favor the formation of the para-isomer in homogeneous conditions. The ortho/para ratio is determined by the electron densities at the ortho- and para-positions in the phenoxy radical; this is not the controlling factor within zeolites. Mechanistic details merit clarification in order to discriminate between the intra- and intermolecular pathways.[27,36,37] From an experimental point of view, the presence of phenol in the reaction medium is indicative of inter-molecular acylation (Scheme 5.12a) or deacylation (Scheme 5.12b).

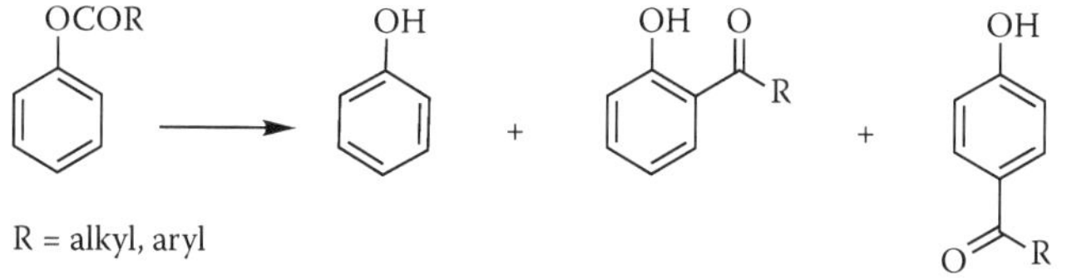

R = alkyl, aryl

Scheme 5.11

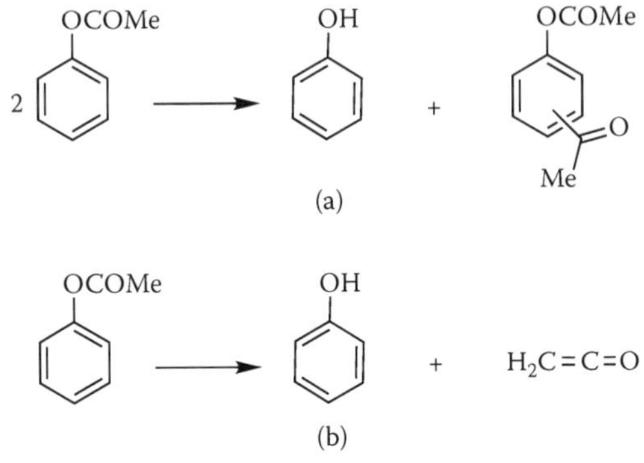

Scheme 5.12

While the first process represents a positive event because acetoxyaceto-phenones are convertible into HAPs and can give further intermolecular phenol acylation affording both *ortho*- and *para*-HAP, the second process produces ketene, which, being highly reactive, represents the most important source of coke responsible for heterogeneous catalyst deactivation.

5.2.1   *Metal triflates*

The Fries rearrangement of phenyl- and 1-naphthyl esters can be efficiently performed in the presence of hafnium triflate.[10] The method is based on the one previously described for the Friedel–Crafts acylation of arenes with acyl chlorides.[38,39] The reaction occurs in toluene–nitromethane mixtures at 100°C for 6 h. Several examples of this Fries isomerization with synthetic application are reported in Table 5.5. In all cases, complete regioselectivity is obtained, and 2-acylated phenol or naphthol derivatives are isolated in good yields.

Results of the Fries rearrangement of 1-naphthyl esters pro-moted by scandium triflate (5% mol) are reported in Table 5.6.[12] The reaction proceeds smoothly in toluene at 100°C for 6 h to produce 1-hydroxy-2-naphthyl ketones in high selectivity. The yields obtained are much higher than those achieved in the stoichiometric aluminum-chloride-promoted reactions.[40]

These results confirm that the Fries rearrangement can be performed in the presence of specific transition metal triflates as homogeneous, highly active catalysts with low toxicity, moisture and air tolerance. However, the high cost of these catalysts limits their use to small-scale production.

Bismuth(III) triflate tetrahydrate can be utilized as a more commer-cially available catalyst for the Fries rearrangement of PAs to 2-hydrox-yaryl ketones.[41] In the reaction with 1-naphthyl acetate, bismuth triflate (10% mol) gives the best yield in 2-acetyl-1-naphthol, the amount of

Table 5.5 Hafnium-triflate-promoted Fries rearrangement of phenyl esters

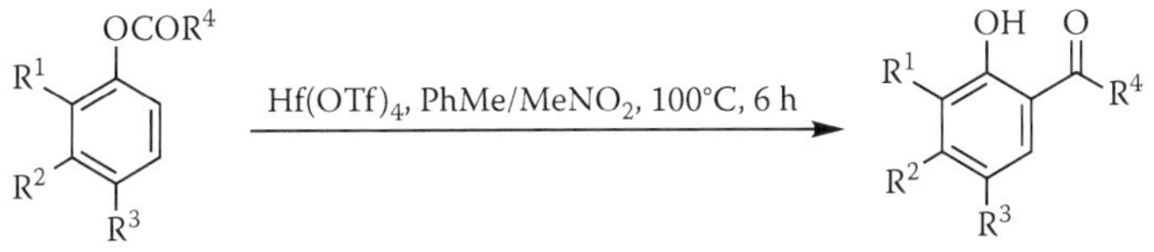

OCOR⁴, R¹, R², R³ → Hf(OTf)₄, PhMe/MeNO₂, 100°C, 6 h → OH O, R¹, R², R³, R⁴

R¹	R²	R³	R⁴	Catalyst (mol %)	Yield (%)
H	OMe	H	Me	20	53
H	OMe	H	$c\text{-}C_6H_{11}$	20	72
H	Me	H	Me	20	64
H	Buᵗ	H	Me	10	60
H	H	Me	Me	20	62
CH=CH-CH=CH	H	Me	5	76
CH=CH-CH=CH	H	$c\text{-}C_6H_{11}$	5	82

Table 5.6 Scandium-triflate-promoted Fries rearrangement of naphthyl esters

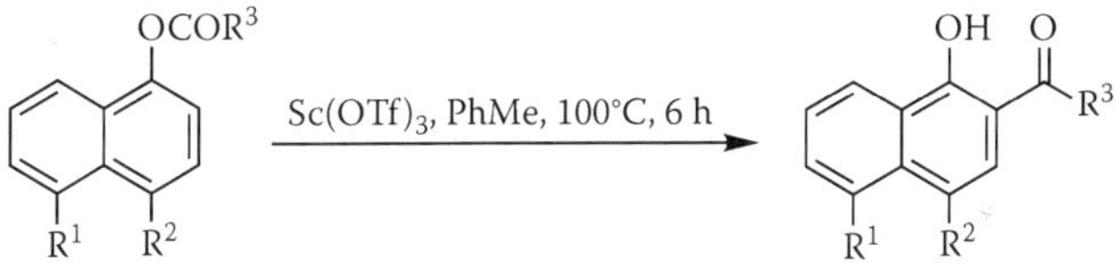

OCOR³, R¹, R² → Sc(OTf)₃, PhMe, 100°C, 6 h → OH O, R³, R¹, R²

R¹	R²	R³	Yield (%)
H	H	Me	85
H	H	Et	89
H	H	C_5H_{11}	85
H	H	$c\text{-}C_6H_{11}$	78
H	OMe	Me	80
OMe	H	Me	66

1-naphthol being minimized. Synthetic results reported in Table 5.7 show that 2-acetyl-1-naphthol derivatives are efficiently obtained in toluene or nitroethane at 110°C for 3–15 h.

5.2.2   Metal halides

Polyhydroxybenzophenones can be synthesized in high yield by the Fries rearrangement of polyhydroxybenzoates in the presence of zirconium(IV), iron(III), tin(IV), and bismuth(III) halides (20%–200%) at 30°C–90°C.[42]

Table 5.7 Bismuth(III)-triflate-promoted Fries
rearrangement of naphthyl esters

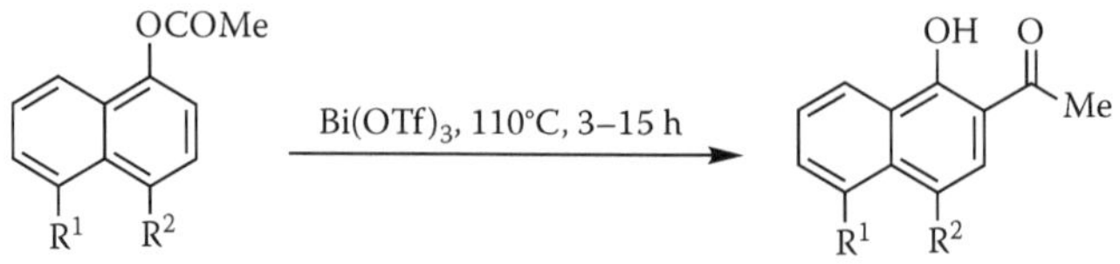

R¹	R²	Solvent	Yield (%)
H	H	PhMe	80
H	H	EtNO$_2$	84
H	OMe	PhMe	71
H	OMe	EtNO$_2$	81
OMe	H	PhMe	65
OMe	H	EtNO$_2$	66
H	Cl	PhMe	82
H	Cl	EtNO$_2$	41
H	O$_2$CMe	PhMe	64
H	O$_2$CMe	EtNO$_2$	6

Different acid catalysts, including liquid hydrofluoric acid, are also utilized for the same purpose.[43,44]

A Fries-like rearrangement is employed in the preparation of 6-acyl-2(3H)-benzoxazolones (**9**, [Table 5.8], X = O) and 6-acyl-2(3H)-benzothiazolones (**9**, [Table 5.8], X = S).[45] The rearrangement is performed by heating at 165°C a mixture of the easily synthesizable N-acylated derivative **8** with aluminum chloride for 3 h. Products **9** with a variety of substituents on the acyl moiety can be prepared in high yield (Table 5.8).

1-Butyl-3-methylimidazolium chloroaluminate ([mim]⁺Al$_2$Cl$_7$⁻) can be utilized as a solvent as well as a Lewis acid catalyst in the Fries rearrangement reaction of PBs.[46] In chloroaluminate ionic melts, the Lewis acid species present in aluminum-chloride-rich compositions is well established and is known to be Al$_2$Cl$_7$⁻.[47] The composition of the ionic liquid is usually expressed as the apparent aluminum chloride mole fraction, N. Accordingly, they are classified as basic, neutral, or acidic melts where N is 0.00–0.50, 0.50, and 0.50–0.67, respectively. The Fries rearrangement reaction of PBs is carried out with a melt showing $N = 0.67$. Rearrangement of PB in the presence of anisole provides evidence for intermolecularity of the process. The reaction gives good yields of the rearrangement products in the case of substrates bearing electron-donating groups, whereas with substrates bearing electron-withdrawing groups, considerable debenzoylation is observed (Table 5.9).

Table 5.8 Aluminum-chloride-promoted Fries rearrangement of 6-acyl-2(3H)-benzoxazolones and 6-acyl-2(3H)-benzothiazolones

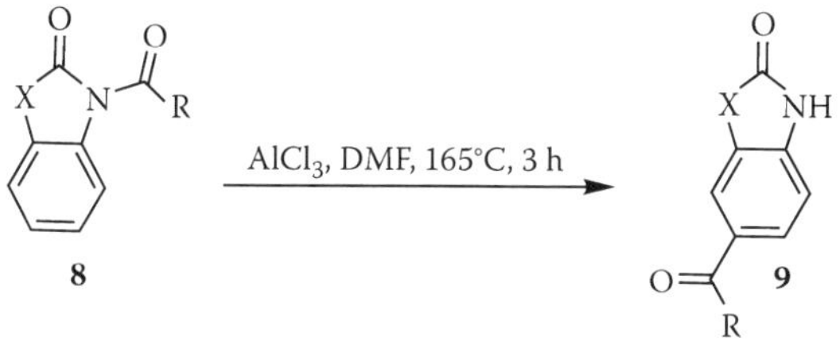

X	R	Yield (%)
O	Me	81
O	Et	89
O	Ph	84
O	3-FC$_6$H$_4$	82
O	4-FC$_6$H$_4$	83
O	4-ClC$_6$H$_4$	92
O	4-NO$_2$C$_6$H$_4$	85
S	Me	81
S	Et	86
S	3-FC$_6$H$_4$	94
S	4-FC$_6$H$_4$	84
S	4-ClC$_6$H$_4$	90
S	4-NO$_2$C$_6$H$_4$	88

5.2.3 Metals

An interesting and efficient Fries rearrangement is based on the use of zinc powder as a nontoxic and inexpensive catalyst coupled with microwave irradiation.[48] A variety of O-acylated phenols undergo Fries rearrangement selectively at one position, giving a single isomer in quite good yield. In some cases, different isomers are formed depending on the use of microwave or oil-bath heating (Table 5.10).

Thus, in the case of PA, *ortho*-HAP is obtained in 75% isolated yield when irradiated in a microwave oven for 3 min, whereas, when the reaction is carried out with the same quantities in an oil-bath at 60°C for 6 h, *para*-HAP is obtained in 70% isolated yield. Under microwave irradiation, para-migration takes place, whereas, under oil-bath conditions, ortho-migration is observed. Moreover, 1-naphthyl- and 2-naphthyl acetates undergo the Fries rearrangement at the ortho-position in 85% and 90% isolated yields, respectively, when subjected to microwave irradiation in the presence of zinc powder and dimethylformamide for 10 and 8 min,

Table 5.9 Fries rearrangement of phenyl benzoates in the presence of $[mim]^+Al_2Cl_7^-$ at 120°C for 2 h

Substrate	Yield (%)	Product distribution (%)		
		Ortho	Para	Debenzoylation
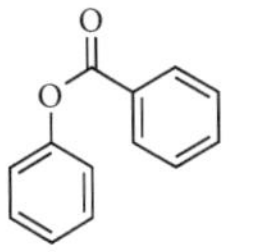 phenyl benzoate	100	17	80	3
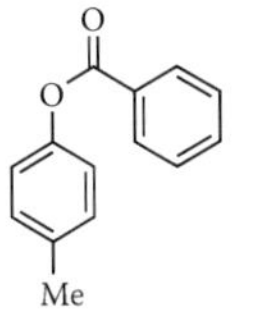 4-methylphenyl benzoate (Me)	100	96	0	4
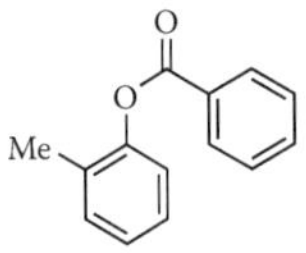 2-methylphenyl benzoate (Me)	100	4	91	5
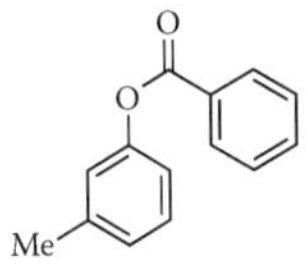 3-methylphenyl benzoate (Me)	100	16	80	4
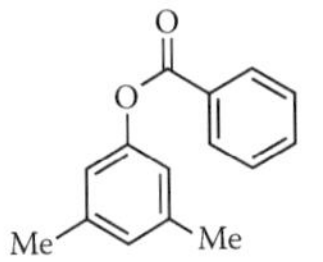 3,5-dimethylphenyl benzoate (Me, Me)	100	94	0	6
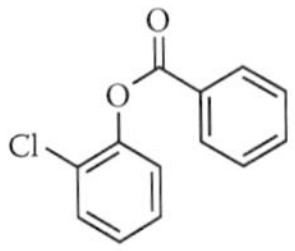 2-chlorophenyl benzoate (Cl)	100	4	76	20
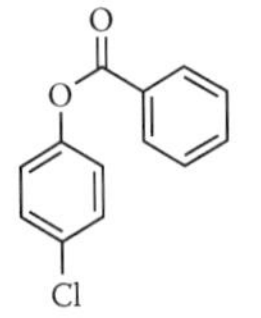 4-chlorophenyl benzoate (Cl)	73	24	0	76
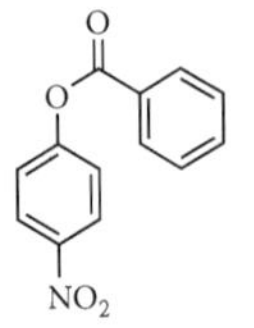 4-nitrophenyl benzoate (NO$_2$)	100	1	0	99

Table 5.10 Zinc-promoted Fries rearrangement of phenyl acetates in dimethylformamide under microwave and oil-bath heating

Substrate	MW					Oil bath				
				Product distribution					Product distribution	
	T (°C)	t (min)	Yield (%)	Ortho	Para	T (°C)	t (min)	Yield (%)	Ortho	Para
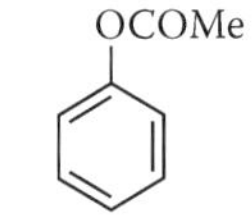 Phenyl acetate (OCOMe)	65	3	75	100		65	360	70		100
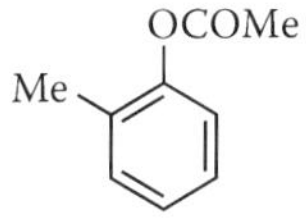 2-Me-phenyl acetate (OCOMe, Me ortho)	68	5	74		100	68	30	69	100	
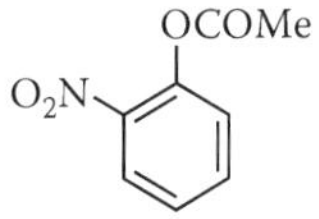 2-NO₂-phenyl acetate (OCOMe, O₂N)	64	13	76	100		64	60	65		100
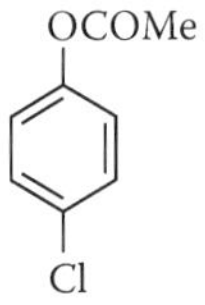 4-Cl-phenyl acetate (OCOMe, Cl)	68	18	79	100		68	120	75	100	
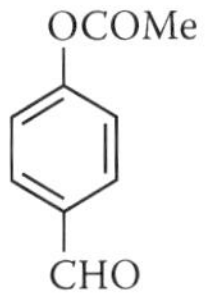 4-CHO-phenyl acetate (OCOMe, CHO)	46	10	77	100		46	90	70	100	
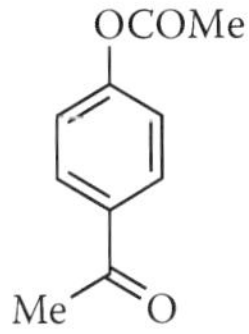 4-acetyl-phenyl acetate (OCOMe, MeCO)	49	12	70	100		49	75	68	100	
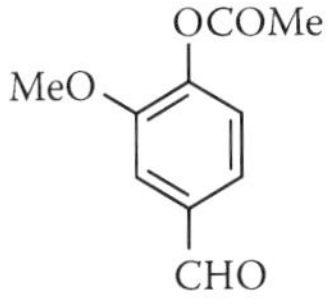 2-MeO-5-CHO-phenyl acetate (OCOMe, MeO, CHO)	84	25	76	100		84	60	75	100	

respectively. However, deacetylation takes place when the same reaction is carried out using an oil-bath. It is confirmed that zinc powder is the real catalyst, and zinc oxide, possibly produced in situ, is not active. The zinc powder can be recovered and reused after washing with diethyl ether and diluted hydrochloric acid, giving, in the case of PA, the same results for at least four times.

5.2.4   sec-Butyllithium

Ortho-specific metal-promoted Fries rearrangement can be achieved, starting from *ortho*-bromophenyl esters and *sec*-butyllithium.[49] The reaction is performed at −95°C with 4:1:1 tetrahydrofuran:diethyl ether:hexane ratio followed by stirring for 30 min, followed by an additional 30 min at 78°C. The *ortho*-hydroxyaryl ketones are the sole isomers obtained, and the regioisomeric *para*-hydroxyaryl ketones are not obtained (Table 5.11).

Comparative experiments show that the acyl migration step is an intramolecular process. Interestingly, it is shown that pivaloates **10**, having the ester functionality separated from the aromatic nucleus by a carbon chain, can participate in this acyl migration reaction (Table 5.12).

Table 5.11 *sec*-Butyllithium-promoted Fries rearrangement of *ortho*-bromophenyl esters

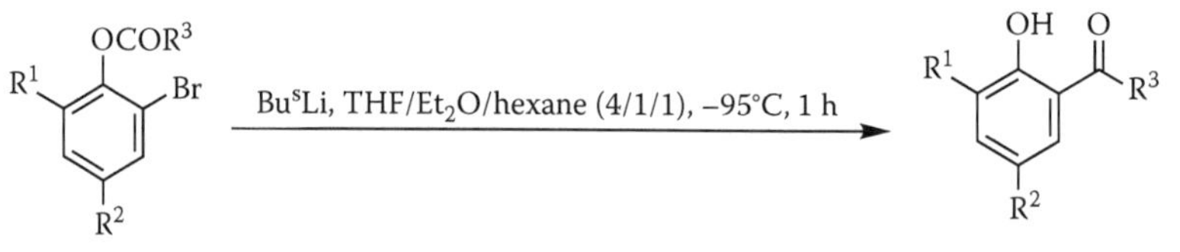

R¹	R²	R³	Yield (%)
H	H	Et	17
H	H	Pr_i	62
H	H	Bu^t	76
H	H	Ph	7
H	H	1-adamantyl	7
H	H	Bu^tCH_2	91
H	Me	Bu^t	71
H	Me	Bu^tCH_2	84
H	H	$C(Me)_2CH_2Cl$	49
Me	Me	Et	31
Me	Me	Pr^i	72
Bu^t	Bu^t	Me	52

Table 5.12 *sec*-Butyllithium-promoted Fries rearrangement of different pivaloates

Bu^{t}OCO—(CH$_2$)$_n$, Br, aryl **10**

Bu^{s}Li, THF/Et$_2$O/hexane (4/1/1), −95°C, 1 h

Bu^{t}OCO—(CH$_2$)$_n$, Li, aryl **11**

LiO—C(Bu^t)(O−(CH$_2$)$_n$)— aryl **12**

Bu^t—C(=O)— aryl —LiO(CH$_2$)$_n$ **13**

HO—C(Bu^t)(O−(CH$_2$)$_n$)— aryl **14**

Bu^t—C(=O)— aryl —(CH$_2$)$_n$OH **15**

		Product distribution (%)	
n	Yield (%)	14	15
1	86	100	0
2	67	50	50
3	58	0	100
4	16	0	100

Baldwin's rules[50] predict that exo-trig cyclization such as **11** → **12** (in equilibrium with the open form **13**) is favored when n = 0–3. Accordingly, this rearrangement occurs readily for these substrates. Interestingly, the pivaloate **10** (n = 1) furnishes only compound **14** (n = 1) in 86% yield; conversely, the same process with n = 2 provides an approximately equal mixture of isomers **14** and **15** in 77% yield, whereas when n = 3, only the hydroxyketone **15** is isolated in 58% yield.

Anionic ortho-Fries rearrangement is performed by treatment of variously substituted aryl carbamates with *sec*-butyllithium at −78°C in tetramethylethylenediamine–tetrahydrofuran solution for 8–12 h.[51] The reaction involves the selective ortho-metalation of the aryl groups. The meta-cooperative metalation effect is responsible for the production of some extremely hindered aromatic compounds (Table 5.13).

Table 5.13 *sec*-Butyllithium-promoted Fries rearrangement of phenyl carbamates

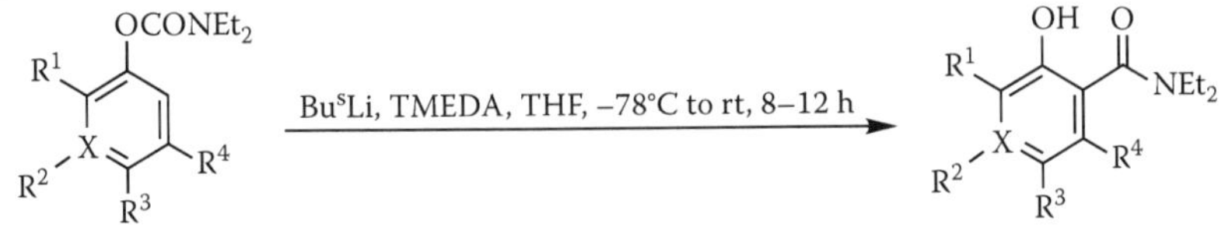

R¹	R²	R³	R⁴	X	Yield (%)
H	H	H	H	C	75
H	H	Me	H	C	70
COOH	H	Cl	H	C	60
CONEt$_2$	OH	H	H	C	75
OMe	H	H	H	C	68
OCH$_2$O		H	H	C	58
H	H	H	OCONEt$_2$	C	86
H	H	H	Cl	C	71
H	—	H	H	N	40

The double 1,4-carbamoyl rearrangement to a hydroquinone diamide is performed in 25% yield. The Fries rearrangement is also observed in good yields in the naphthyl-, phenanthryl-, pyridyl-, and quinolinyl carbamate series.

5.2.5   Zeolites

In the Fries liquid-phase rearrangement of PA carried out in the presence of ZSM-5(41) zeolite at 170°C, a good shape selectivity (ortho/para = 0.17) can be reached after 24 h by keeping modest the conversion of the starting ester (~20%).[52] When the reaction is repeated over Nu-10 zeolite, a 1:1 mixture of ortho- and para-products is isolated; this different behavior can be ascribed to the smaller pore dimension of Nu-10 zeolite, which hampers the entrance of the reagents producing both isomers.

A comparative study shows that Nafion and ZSM-5 give similar activity but, in contrast, ZSM-5 affords higher selectivity in favor of para-substituted products.

The same reaction was performed at 200°C for 24 h in the presence of ZSM-5(40), giving a 69% PA conversion with relatively low formation of phenol and by-products, and an unexciting selectivity toward the para product (ortho/para = 0.67).[53] Treatment of the catalyst with triphenylchlorosilane, which deactivates its outer surface, shows an appreciable effect on some catalysts' properties, such as performance as well as ortho/para ratio, which are somewhat higher than those of the starting catalyst (81% conversion, ortho/para = 0.5). This suggests that the reaction mainly occurs in the pores of ZSM-5 material. A catalyst deactivation is observed after 20 h (45%–47% conversion); in a continuous-flow liquid-phase reaction at the same temperature, the catalyst undergoes a lower deactivation rate, clearly attributable to the shorter contact time between the catalyst and the products due to the washing effect, which can remove deactivating species from the catalyst surface; however, a modest selectivity is observed (ortho/para = 0.91). A poor performance of the same catalyst in agreement with previous studies[33,37] is observed in continuous gas-phase experiments carried out at 420°C.

ZSM-5(>20) zeolite in the pure form or as agglomerates joined with alumina and silica was studied as catalyst in the same gas-phase reaction.[54] Low temperatures favor adsorption of PA on the catalyst, whereas at higher ones, the formation of undesired by-products is increased. The best results are achieved by carrying out the reaction on pure ZSM-5 at 265°C, which shows 86% conversion accompanied by a very high ortho-selectivity (ortho/para = 35.5) and only 27% phenol production. On the contrary, the performance of the same catalyst containing ~40% of binder worsens slightly with alumina (75% conversion, 18.2 ortho/para ratio, 23% phenol) and more dramatically with silica (44% conversion, 5.4 ortho/para ratio, 63% phenol), evidencing the effect of a delicate balance between Lewis and Brönsted acid sites. The FT-IR analyses[55] of the catalysts show significant differences from other ZSM-5-type zeolites, with probably a partial silylation of the external surface, resulting in the production of a large amount of very weak acid external silanols and an increase in the hydrophobic character. ACD/3D program calculations permit the conclusion that the ortho-isomer is slightly compatible with the porosity of ZSM-5, in which the formation of the para-isomer can be favored.[56] Thus, the high level of ortho-selectivity observed allows the hypothesis that, with the present catalyst, the reaction occurs on the external surface. The partial silylation of the external surface increases its hydrophobic character with a pronounced decrease in the amount of Lewis acid sites and promotes the adsorption of both ester feed and ketones formed while decreasing that of the polar phenol, thus reducing the formation of the para-product.[57]

GaZSM-5(60) can be employed as catalyst to promote the same reaction in the gas phase at 250°C using a tubular downflow pyrex reactor.[58] The conversion of PA is 71%, the *ortho*-HAP is produced in 46% yield, and the *para*-HAP yield is <5%.

The influence of the solvent on the rate and selectivity of PA transformation over BEA was deeply investigated.[59] Large differences are observed in both catalyst activity and product distribution by carrying out the reaction in sulfolane or dodecane. Sulfolane (E_T^N = 0.410) significantly favors the formation of the para-product with respect to dodecane (E_T^N = 0.012), and the production of *ortho*-HAP is seven times slower in sulfolane than in dodecane. As already stated, the reaction orders are likely due to differences in adsorption of the solvents on the acid sites: sulfolane can compete with PA and the reaction products for adsorption on the acid sites, whereas the adsorption of dodecane can be considered as negligible. Of consequence, sulfolane also limits catalyst deactivation, as already observed in the direct acylation of aryl ethers.[60]

Regarding catalyst deactivation, it is also found that the phenolic products (phenol, *ortho*-HAP, and *para*-HAP) are strong inhibitors of PA conversion.[21,61] Whereas all the phenolic compounds contribute to catalyst inhibition during the initial reaction period, phenol is replaced by competitive adsorption of *ortho*-HAP and *para*-HAP after a longer reaction time, when they are formed to a greater extent. When adding fresh PA, they are extracted from the catalyst so that the catalyst activity is mostly regained. Deactivation by coke[53,57] only plays a role after prolonged reaction times. Zeolite deactivation can be lowered by performing the Fries reaction in the trickle bed reactor (Figure 5.2) already utilized in anisole acylation.[62] In the initial period, the conversion in the batch reactor is higher than in the trickle bed reactor (15% versus 8% conversion after 30 min); this is probably caused by the higher heat capacity of the latter. Moreover, after 180 min, with the trickle bed reactor, the conversion exceeds that in the batch reactor (27% versus 14%). In addition, the higher the catalyst:reactant ratio, the higher the conversion.

The modification of BEA zeolite by surface deposition of silica and impregnation with cerium oxide was studied as a tool to improve the selectivity of the reaction.[63] The number of acid sites, particularly the strong ones, on BEA zeolite decreases with increasing amounts of silica deposited on its surface. Moreover, there is no severe pore blocking after deposition. On the contrary, cerium oxide impregnation affords a catalyst with decreased adsorption capacity because part of the cerium oxide is deposited in the channels of the zeolite crystals and blocks the porous system. In addition, cerium oxide modification creates new weak acid sites on the zeolite surface. Silica modification decreases catalytic activity but slightly increases selectivity with respect to all *ortho*-HAP, *para*-HAP and *para*-acetoxyacetophenone, in comparison to the unmodified BEA zeolite, and the stability of the catalyst is also improved after modification. The best reaction results are obtained over 16% cerium-oxide-modified catalyst, the selectivity with respect to the C-acetylated products being increased to about 70% while the conversion remains 60%–80%.

5.2.6   Heteropoly acids (HPAs)

HPA- and HPA-salts-catalyzed Fries rearrangement of phenyl esters have been deeply investigated.[64–68] Three kinds of catalysts, namely, bulk HPA, silica-supported HPA (HPA/SiO$_2$)[69] (including its cesium salts), and sol-gel silica-supported HPA (HPA-SiO$_2$),[70,71] were compared. The insoluble

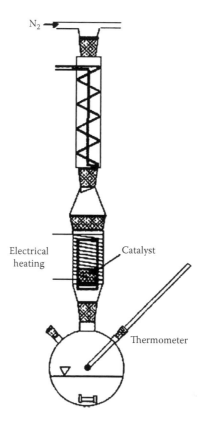

Figure 5.2 Trickle-bed reactor for the Fries rearrangement of phenyl acetate in the presence of BEA zeolite. (From Freese, U., Heinrich, F., and Roessner, F., *Catal. Today*, 49, 237, 1999. With permission.)

salt $Cs_{2.5}H_{0.5}PW$ (CsHPA) catalyzes the reaction in polar media such as nitrobenzene with 9% PA conversion and with a turnover frequency (TOF) of 15 min^{-1}; although it is less active per unit weight than the homogeneous HPA or HPA/SiO$_2$, it is also more selective to HAPs, making less phenol. The explanation of this behavior may be that the less hydrophilic CsHPA[72] possesses stronger proton sites than HPA or HPA/SiO$_2$. Some leaching of HPA from HPA/SiO$_2$ (~10% per run) is observed during recycling studies, probably due to the ready solubility of HPA in PA at elevated temperatures.[65,66] These observations reopen the question of the true heterogeneity of the catalyst. Much better reusability shows the more stable CsHPA that is insoluble, and in this case, the reaction is truly heterogeneous. In addition, the low-porous CsHPA makes much less coke (~2%) than the high-porous HPA-SiO$_2$; the amount of coke remains constant in successive runs. However, CsHPA reuse shows a decline of activity and selectivity to C-acylated products, which indicates loss of catalyst acidity.

Palladium doping (2.1%) of CsHPA (Pd-CsHPA) further improves the catalyst performance upon reuse. The palladium-doped catalyst performed no better than the undoped one, but it can be regenerated with full recovery of both activity and selectivity over five successive runs. Only a small decrease in activity in the first run for Pd-CsHPA compared with undoped CsHPA can be noted, probably because the palladium partly exists as Pd(II) in the first run, whereas in the subsequent runs it mostly exists as Pd(0) due to reduction of Pd(II) by the reaction medium.

5.2.7   Clays

Among the various heterogeneous catalysts studied for performing the Fries rearrangement, clays and modified clays merit some attention.[73] The activity of K10 Montmorillonite (K10), acid-treated Montmorillonite K10 (HK10), and sodium- and aluminum-exchanged K10 Montmorillonite (NaK10 and AlK10) in the Fries rearrangement of phenyl and naphthyl acetates and benzoates was compared. Significant results are reported in Table 5.14. Reactions are performed without solvent by heating the mixture at 140°C for 5 h. In the case of PA, the reaction proceeds very smoothly, with all catalysts giving *para*-HAP exclusively in very high yield. By using a twofold excess of catalyst, an increase in *ortho*-HAP is observed.

Similar good results are achieved with other phenyl and naphthyl esters. The orienting influence is dependent on the migrating group: with a small acyl group, migration inside the clay interlayer to form the thermodynamically more stable para-isomer is facile, but when the catalyst is present in excess, both the nucleophilic centers in the substrate are complexed separately, rendering ion-pair mobility more difficult, thus favoring the formation of the ortho-isomer also. It must be underlined, however, that one drawback of the present method is represented by the great amount of clay catalyst utilized.

5.2.8   Nafion

Nafion (5%) in a refluxing solution of phenyl esters in nitrobenzene promotes the formation of ketone products in good yields (Table 5.15).[74] As shown in Table 5.15, the reaction is of general applicability for phenyl esters of aromatic carboxylic acids. The catalyst can be recovered by filtration and reused after simple regeneration.

Silica composite materials constituted by Nafion entrapped in a highly porous silica matrix (13% and 40% Nafion) are utilized as catalysts for the Fries rearrangement of PA.[57] The conversions of the reactions performed in cumene are 10% and 16%, respectively, and the selectivity with respect to *para*- and *ortho*-HAP ranges from 20% up to 56%, phenol being the main side product accompanied by deposition of carbonaceous materials; these

Table 5.14 Clay-promoted Fries rearrangement of phenyl and
1-naphthyl esters at 140°C for 5 h

Catalyst	Substrate	Conversion (%)	Product distribution (%)		
			2-isomer	4-isomer	phenol/1-naphthol
K10	PA	88	3	97	0
NaK10	PA	81	0	100	0
HK10	PA	100	0	100	0
AlK10	PA	100	0	100	0
K10	PB	90	11	69	20
NaK10	PB	94	26	68	6
HK10	PB	100	9	77	14
AlK10	PB	100	12	74	14
K10	1-Naphthyl acetate	72	94	0	4
NaK10	1-Naphthyl acetate	66	94	0	6
HK10	1-Naphthyl acetate	81	89	0	11
AlK10	1-Naphthyl acetate	61	95	0	5
K10	1-Naphthyl benzoate	43	100	0	0
NaK10	1-Naphthyl benzoate	45	89	0	11
HK10	1-Naphthyl benzoate	73	89	0	11
AlK10	1-Naphthyl benzoate	61	87	0	13

results are comparable to those obtained with USY. By using phenol as solvent instead of cumene, a tremendous change in the chemical pathway is
found: in particular, by using 13% Nafion in the silica composite catalyst,
a selectivity of more than 90% (with respect to the two HAPs) is obtained
with a conversion of about 21%. In the case of BEA, a higher conversion is
observed (41%) but accompanied by lower selectivity (76%). Unfortunately,
it is impossible to reuse the catalysts without reactivation as only traces of
product can be obtained.

5.3   Photo-Fries rearrangement

A series of studies was performed on the Fries rearrangement carried out
under photocatalytic conditions, with the aim of improving the process
from the preparative and environmental points of view. Various lithium-,

Table 5.15 Nafion-promoted Fries rearrangement of PBs
in nitrobenzene at reflux for 12 h

Substrate	Yield (%)	Product distribution (%)	
		Ortho	Para
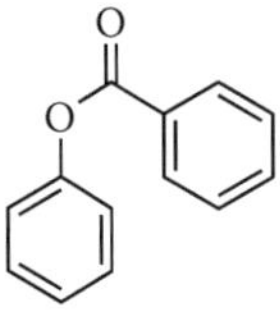	73	34	66
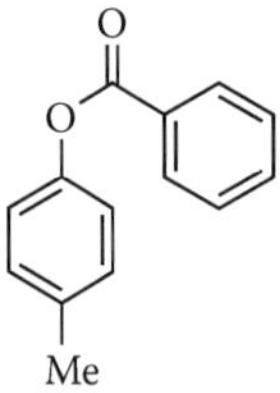	70	100	
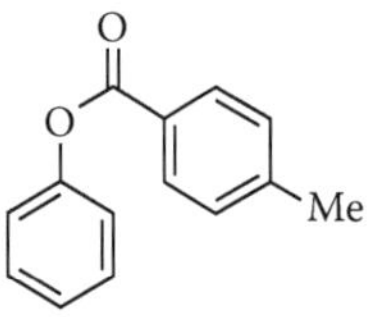	63	29	71
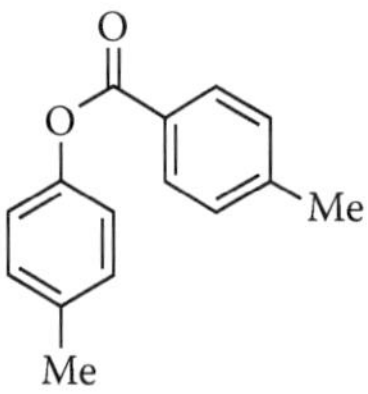	72	100	
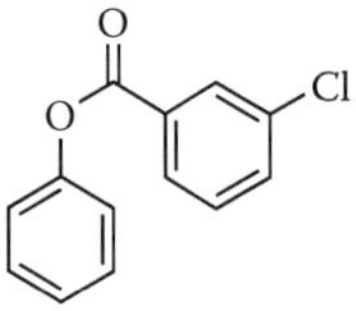	75	28	72
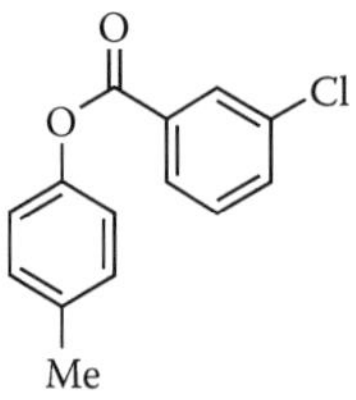	71	100	

sodium-, and potassium-exchanged Y, X, and ZSM-5 zeolites were considered.[75] PA or PB are previously adsorbed on the selected zeolite, and
the zeolite samples are successively irradiated (~2 h) in hexane as a slurry
to about 30% conversion. The photoreaction within LiX, NaX, LiY, and
NaY zeolites gives a small percentage (<10%) of the para-isomer, whereas
within KX and KY zeolites, the ortho-isomer is the exclusive product. This
selectivity is not the result of a shape exclusion because both the ortho-
and para-isomers fit very well within the supercage. On the contrary,
it probably results from the restriction imposed on the mobility of the
phenoxy and acyl fragments by the supercage framework, and from the
interaction between the cations and the two reactive fragments.[76] Although
the shape and size of pentasils are such that only the para-isomer can fit
within the channels of these materials, no selectivity is achieved when the
photolysis of PA included within ZSM-5 is conducted as a hexane slurry
(0.9 ortho/para ratio). Hexane displaces the reactant from the interior of
ZSM-5, and most of the reactants are present in the hexane solvent and
not inside the zeolite, thus giving significant yield of the ortho-isomer.
On the other hand, para-selectivity is achieved by using solvents that do
not displace the included PA from the interior of ZSM-5, in agreement
with results achieved in the same reaction catalyzed by BEA.[59] Because the
internal structure of ZSM-5 is highly hydrophobic, water is not expected
to fill its channels. Similarly, 2,2,4-trimethylpentane solvent molecules,
being too large, are not expected to enter the channels. These solvents
keep both the reactants and the reactive fragments within the channels of
ZSM-5: thus, the reaction is forced to occur inside the channels, showing
high para-selectivity (water: ortho = 6%, para = 62%; 2,2,4-trimethylpentane:
ortho = 5%, para = 63%).

A further example of the crucial role played by solvent polarity in controlling the ortho/para ratio in photo-Fries rearrangement is represented
by the behavior of variously substituted PAs.[77] For example, irradiation of
2,5-dimethylphenylacetate in methanol at 254 nm leads to the formation
of 2-hydroxy-3,6-dimethylacetophenone **16** (Table 5.16, R^1 = R^2 = Me) as
the major product (50% yield), accompanied by 4-hydroxy-2,5-dimethyl-
acetophenone **17** (R^1 = R^2 = Me) (30% yield) and 2,5-dimethylphenol **18**
(R^1 = R^2 = Me) (20% yield). Increasing the polarity and viscosity of the
solvent (2:1 methanol:water) also increases the ortho/para ratio (3.35) while
diminishing the 2,5-dimethylphenol concentration (12%) as a result of the
stronger solvent cage surrounding the radical pair. Similar behavior is
achieved with different phenols. When 2-methoxy-4-methylphenyl acetate
is irradiated in a greatly viscous medium such as glycerol, the ortho/para
ratio of products is enhanced to >30.

Similar interesting results are achieved with naphthyl esters (Scheme
5.13).[78] Indeed, while in solution, both 2-acetyl- (or 2-benzoyl-) naphthol

Table 5.16 Photo-Fries rearrangement of 2,5-disubstituted PAs

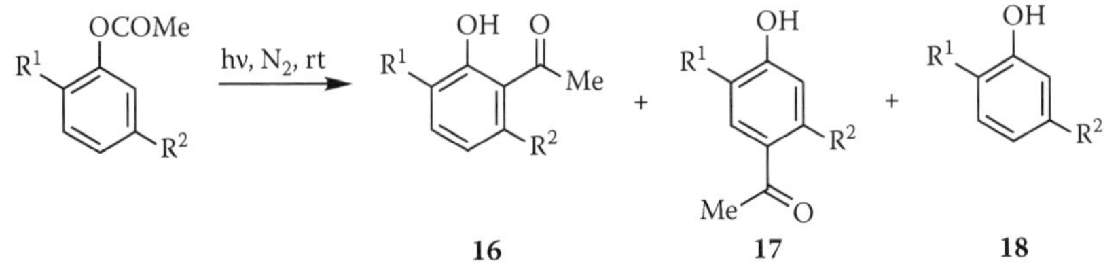

					Product distribution (%)		
			Conversion	Yield			
R^1	R^2	Solvent	(%)	(%)	16	17	18
Me	Me	MeOH	82	79	50	30	20
Me	Me	MeOH:H_2O (2:1)	52	50	68	20	12
Me	Pr^i	MeOH	81	75	48	20	32
Me	Pr^i	MeOH:H_2O (2:1)	91	88	58	16	26
Pr^i	Me	MeOH	83	79	54	30	16
Pr^i	Me	MeOH:H_2O (2:1)	74	71	68	24	8
OMe	Me	MeOH	84	82	46	26	28
OMe	Me	MeOH:H_2O (2:1)	53	51	68	20	12
OMe	Me	Glycerol	72	50	64	2	34

and 4-acetyl- (or 4-benzoyl-) naphthol are formed, within zeolites the acetyl and benzoyl radical seeks only the 2-position of the naphthoxy radical (98% 2-benzoylnaphthol with KY and 96% 2-benzoylnaphthol with NaY in hexane).

Very high ortho-regioselective Fries rearrangement is performed when variously functionalized PAs are subjected to irradiation in hexane and in the presence of solid anhydrous potassium carbonate during 12 h at 25°C (Table 5.17).[79] Under the aforementioned conditions, PAs are converted into the corresponding *ortho*-hydroxyacetophenones **19** in high yields (*para*-hydroxyacetophenones **20** are obtained in lower yield), giving a series of compounds, some of which are not easily accessible through the classical aluminum-chloride-promoted reaction (i.e., 2-hydroxy-5-methoxyacetophenone).

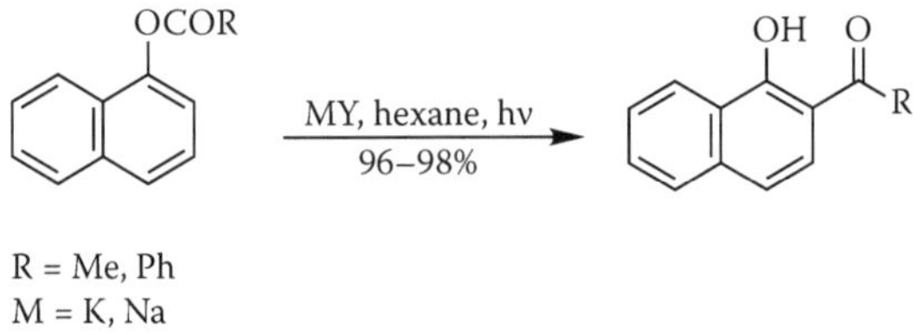

R = Me, Ph
M = K, Na

Scheme 5.13

Table 5.17 Photo-Fries rearrangement of PAs in the presence of potassium carbonate

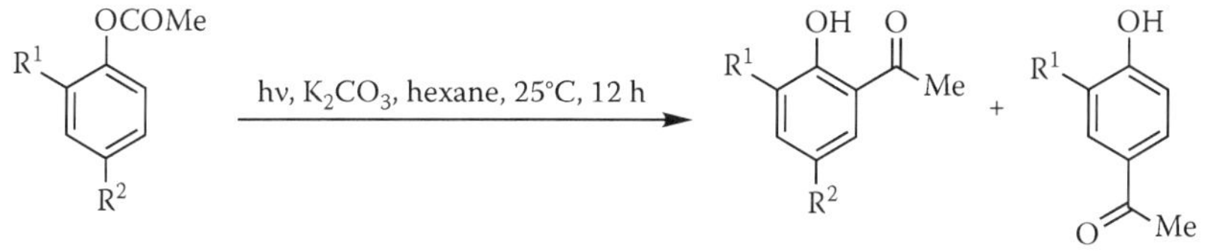

R^1	R^2	Yield (%)	Product distribution (%) 19	20
H	H	89	88	12
Me	H	85	87	13
H	Me	86	100	—
H	OMe	89	100	—
H	Cl	88	100	—
Me	Me	90	100	—

Further detailed studies confirm the impressive role played by the support in the photo-Fries rearrangement of 1-naphthyl esters.[80] Indeed, 1-hydroxy-2-acylnaphthols are obtained in 93%–99% yield upon irradiation (>300 nm) of different naphthyl esters adsorbed on LiY, NaY, and KY.

Phenyl cinnamates **21** undergo photorearrangement for the synthesis of 2′-hydroxychalcones **22** (Table 5.18).[81] The reaction is performed in micellar medium in order to improve yield and selectivity. Indeed, under these conditions, micellar environmental factors such as compartmentalization, localization, preorientation, cage, polarity, and counterion effects can dramatically influence chemical behavior. The micellar solution of the cinnamate is prepared by suspending the reagent in a water solution of sodium dodecyl sulfate (SDS), stirring for 10–12 h and irradiating at 254 nm for 8 h. The reaction gives excellent yields of products **22** accompanied by by-products **23**, in contrast to the analogous photoreaction carried out in organic solvents.[82]

The high yield of products is attributed to the micellar compartmentalization (protection of the excited state from quenching), preorientation effect (molecules located in the interface region with aromatic groups toward the hydrophobic core), and cage effect (restriction of the translational freedom of the substrate, resulting in the freezing of intermediates).

Microwave-induced Fries rearrangement can also be efficiently performed by using an aluminum chloride–zinc chloride mixture supported on silica under solventless conditions.[83] The reaction can be applied to different phenyl and naphthyl acetates. *ortho*-HAP is isolated in 95% yield

Table 5.18 Photo-Fries rearrangement of phenyl cinnamates
promoted by sodium dodecyl sulfate

							Product distribution (%)	
R^1	R^2	R^3	R^4	R^5	t (h)	Yield (%)	**22**	**23**
H	H	H	H	H	6	85	82	18
H	H	Me	H	H	5	88	83	17
H	H	H	Me	H	7	90	78	22
H	OH	H	H	H	6	70	100	0
OH	OH	H	H	H	8	75	100	0
OH	OH	H	OMe	OMe	8	75	100	0

with 100% selectivity. It is worth noting that the reaction does not proceed
on zinc chloride or aluminum chloride supported alone on silica gel.

An elegant application of this procedure is represented by the Fries
rearrangement of the cinnamyl esters of phenols and naphthols **24**. In
fact, when using these substrates, the ortho-rearranged product chalcones spontaneously cyclize by intramolecular conjugate addition of the
phenolic hydroxy group to give flavones **25** in 70%–90% isolated yields
(Table 5.19). To demonstrate the efficiency of the methodology, α-naphthyl
acetate mixed with the support is heated at 300°C for 7 min, and the corresponding product is isolated in only 10% yield compared to the 95% yield
gained under microwave irradiation.

Table 5.19 Microwave-induced Fries rearrangement of phenyl cinnamates promoted by aluminum chloride–zinc chloride mixture supported on silica

AlCl$_3$-ZnCl$_2$/SiO$_2$, MW, 7 min

24   **25**

R^1	R^2	R^3	Yield (%)
H	H	Me	87
H	H	OMe	73
CH=CH-CH=CH		H	85
H	CH=CH-CH=CH		79

References

1. Neves, I., Jayat, F., Magnoux, P., Pérot, G., Riberito, F. R., Guberlmann, M., and Guisnet, M. 1994. Acylation of phenol with acetic acid over a HZSM5 zeolite, reaction scheme. *J. Mol. Catal.* 93: 169–179.
2. Jackman, L. M., Petrei, M. M., and Smith, B. D. 1991. Degenerate transesterification of 3,5-dimethylphenolate/3,5-dimethylphenyl esters in weakly polar, aprotic solvents. Reactions of aggregates and complex-induced proximity effects. *J. Am. Chem. Soc.* 113: 3451–3458.
3. Sartori, G., Casnati, G., Bigi, F., and Predieri, G. 1990. Ortho-coordinated acylation on phenol systems. *J. Org. Chem.* 55: 4371–4377.
4. Chapman, N. B. and Shorter, J. 1978. *Correlation analysis in chemistry.* New York: Plenum Press, 357–396.
5. Bigi, F., Casiraghi, G., Casnati, G., Marchesi, S., Sartori, G., and Vignali, C. 1984. Unusual Friedel–Crafts reactions, IX. One-step ortho-acylation of phenols with α,β-unsaturated acyl chlorides. Synthesis of 2'-hydroxychalcones and sorbicillin analogues. *Tetrahedron* 40: 4081–4084.
6. Sartori, G., Casnati, G., Bigi, F., and Bonini, G. 1988. Metal template ortho acylation of phenols: direct synthesis of salicylic acid chlorides and derivatives. *Synthesis* 763–766.
7. Bigi, F., Maggi, R., Sartori, G., and Casnati, G. 1992. Template Houben–Hoesch reaction on metal phenolates. Synthesis of aromatic ketones, nitriles and amides. Crystal structure of dichloro-[2-(1-imino-2,2,2-trichloroethyl)-4-methoxy-phenoxido-*O,N*]boron. *Gazz. Chim. Ital.* 122: 283–289.
8. Anjaneyulu, A. S. R., Mallavadhani, U. V., Venkateswarlu, Y., and Ramaprasad, A. V. 1987. Reaction of acetic anhydride/zinc chloride reagent with phenols: improved yields of hydroxyacetophenones. *Ind. J. Chem.* 26B: 823–826.

9. Kobayashi, S., Moriwaki, M., and Hachiya, I. 1995. 2-Acylation reactions of phenol and 1-naphthol derivatives using Sc(OTf)$_3$ as a Lewis acid catalyst. *Synlett* 1153–1154.

10. Kobayashi, S., Moriwaki, M., and Hachiya, I. 1996. Hafnium trifluoro-methanesulfonate (Hf(OTf)$_4$) as an efficient catalyst in the Fries rearrangement and direct acylation of phenol and naphthol derivatives. *Tetrahedron Lett.* 37: 2053–2056.

11. Kobayashi, S., Moriwaki, M., and Hachiya, I. 1996. Catalytic direct C-acylation of phenol and naphthol derivatives using carboxylic acids as acylating reagents. *Tetrahedron Lett.* 37: 4183–4186.

12. Kobayashi, S., Moriwaki, M., and Hachiya, I. 1997. The catalytic Fries rearrangement and o-acylation reactions using group 3 and 4 metal triflates as catalysts. *Bull. Chem. Soc. Jpn.* 70: 267–273.

13. Neves, I., Jayat, F., Magnoux, P., Pérot, G., Ribeiro, F. R., Gubelmann, M., and Guisnet, M. 1994. Phenol acylation: unexpected improvement of the selectivity to *o*-hydroxyacetophenone by passivation of the external acid sites of HZSM5. *J. Chem. Soc., Chem. Commun.* 717–718.

14. Wang, Q. L., Torrealba, M., Giannetto, G., Guisnet, M., Pérot, G., Cahoreau, M., and Casso, J. 1990. Dealumination of Y zeolite with ammonium hexa-fluorosilicate: A SIMS-XPS study of the aluminum distribution. *Zeolites* 10: 703–706.

15. Padró, C. L. and Apesteguía, C. R. 2004. Gas-phase synthesis of hydroxyac-etophenones by acylation of phenol with acetic acid. *J. Catal.* 226: 308–320.

16. Subba Rao, Y. V., Kulkarni, S. J., Subrahmanyam, M., and Rama Rao, A. V. 1995. An improved acylation of phenol over modified ZSM-5 catalysts. *Appl. Catal. A: Gen.* 133: LI–L6.

17. Chaube, V. D., Moreau, P., Finiels, A., Ramaswamy, A. V., and Singh, A. P. 2002. A novel single step selective synthesis of 4-hydroxybenzophenone (4-HBP) using zeolite H-Beta. *Catal. Lett.* 79: 89–94.

18. Bhattacharyya, K. G., Talukdar, A. K., Das, P., and Sivasanker, S. 2001. Acetylation of phenol with Al-MCM-41. *Catal. Commun.* 2: 105–111.

19. Chaube, V. D., Moreau, P., Finiels, A., Ramaswamy, A. V., and Singh, A. P. 2001. Propionylation of phenol to 4-hydroxypropiophenone over zeolite H-Beta. *J. Mol. Catal. A: Chem.* 174: 255–264.

20. Subba Rao, Y. V., Kulkarni, S. J., Subrahmanyam, M., and Rama Rao, A. V. 1993. A novel acylative cyclization reaction of phenol over modified Y zeolites. *J. Chem. Soc., Chem. Commun.* 1456–1457.

21. Rohan, D., Canaff, C., Magnoux, P., and Guisnet, M. 1998. Origin of the deactivation of HBea zeolites during the acylation of phenol with phenylacetate. *J. Mol. Catal. A: Chem.* 129: 69–78.

22. Borsodi, V. K. 1970. Procedé de preparation de la 2,4-dihydroxybenzo-phénone. FR Patent 2,034,934.

23. Slagan, P. 1977. Preparation of 2,4-dihydroxybenzophenone from resorcinol and benzotrichloride. ZA Patent 7,602,667.

24. Beau, J. P. 1997. Process for the manufacture of hydroxyalkoxybenzo-phenones. U.S. Patent 5,629,453.

25. Dobratz, E. H. and Kolka, A. J. 1968. Production of benzophenones. U.S. Patent 3,403,183.

26. Whelen, M. S. 1968. Process for the preparation of 2,4-dihydroxybenzo-phenone. U.S. Patent 3,371,119.

27. Hoefnagel, A. J. and van Bekkum, H. 1993. Direct Fries reaction of resorcinol with benzoic acids catalyzed by zeolite H-Beta. *Appl. Catal. A: Gen.* 97: 87–102.

28. Jansen, J. C., Creyghton, E. J., Njo, S. L., van Koningsveld, H., and van Bekkum, H. 1997. On the remarkable behaviour of zeolite Beta in acid catalysis. *Catal. Today* 38: 205–212.

29. Paul, V., Sudalai, A., Daniel, T., and Srinivasan, K. V. 1994. HZSM-5 catalysed regiospecific benzoylation of activated aromatic compounds. *Tetrahedron Lett.* 35: 2601–2602.

30. Farkas, J., Békássy, S., Ágai, B., Hegedüs, M., and Figueras, F. 2000. Acylation of resorcinol on clay catalysts. *Synth. Commun.* 30: 2479–2485.

31. Békássy, S., Farkas, J., Ágai, B., and Figueras, F. 2000. Selectivity of *C*-versus *O*-acylation of diphenols by clay catalysts. I. Acylation of resorcinol with phenylacetyl chloride. *Top. Catal.* 13: 287–290.

32. Terunori, F., Masaharu, I., Kazunori, T., and Kenji, S. 1985. Production of acylphenols. JP Patent 60,252,436.

33. Bolognini, M., Cavani, F., Cimini, M., Dal Pozzo, L., Maselli, L., Venerito, D., Pizzoli, F., and Veronesi, G. 2004. An environmentally friendly synthesis of 2,4-dihydroxybenzophenone by the single-step O-mono-benzoylation of 1,3-dihydroxybenzene (resorcinol) and Fries rearrangement of intermediate resorcinol monobenzoate: the activity of acid-treated montmorillonite clay catalysts. *C. R. Chim.* 7: 143–150.

34. Sharghi, H., Hosseini-Sarvari, M., and Eskandari, R. 2006. Direct acylation of phenol derivatives in a mixture of graphite and methanesulfonic acid. *Synthesis* 2047–2052.

35. Gerecs, A. 1964. *The Fries reaction In Friedel–Crafts and related reactions*, ed. G. A. Olah, Vol III, part 1, p. 499. London: John Wiley & Sons.

36. Pouilloux, Y., Gnep, N. S., Magnoux, P., and Pérot, G. 1987. Zeolite-catalyzed rearrangement of phenyl acetate. *J. Mol. Catal.* 40: 231–233.

37. Pouilloux, Y., Bodibo, J.-P., Nebes, I., Gubelmann, M., Pérot, G., and Giusnet, M. 1991. Mechanism of phenylacetate transformation on zeolites. *Stud. Surf. Sci. Catal.* 59: 513–522.

38. Hachiya, I., Moriwaki, M., and Kobayashi, S. 1995. Catalytic Friedel–Crafts acylation reactions using hafnium triflate as a catalyst in lithium perchlorate-nitromethane. *Tetrahedron Lett.* 36: 409–412.

39. Kobayashi, S., Moriwaki, M., and Hachiya, I. 1995. The catalytic Fries rearrangement of acyloxy naphthalenes using scandium trifluoromethane-sulfonate as catalyst. *J. Chem. Soc., Chem. Commun.* 1527–1528.

40. Stoughton, R. W. 1935. The Fries reaction with α-naphthol esters. *J. Am. Chem. Soc.* 57: 202–204.

41. Ollevier, T., Desyroy, V., Asim, M., and Brochu, M.-C. 2004. Bismuth triflate-catalyzed Fries rearrangement of aryl acetates. *Synlett* 2794–2796.

42. Yasuhiko, I., Kazuhiko, H., and Toshiro, O. 1986. Production of polyhydroxy-benzophenone. JP Patent 61,293,945.

43. Shinsuke, F., Masahiro, T., and Sada, K. 1988. Production of 4-(*para*-fluoro-benzoyl)phenols. JP Patent 63,027,454.

44. Mikio, K. and Yuji, M. 1987. Production of 4,4'-(*p*-hydroxybenzoyl)diphenyl ether. JP Patent 62,234,042.

45. Ucar, H., Van derpoorten, K., Depovere, P., Lesieur, D., Isa, M., Masereel, B., Delarge, J., and Poupaert, J. H. 1998. "Fries like" rearrangement: a novel and efficient method for the synthesis of 6-acyl-2(3*H*)-benzoxazolones and 6-acyl-2(3*H*)-benzothiazolones. *Tetrahedron* 54: 1763–1772.
46. Harjani, J. R., Nara, S. J., and Salunkhe, M. M. 2001. Fries rearrangement in ionic melts. *Tetrahedron Lett.* 42: 1979–1981.
47. Hussey, C. L. 1988. Room temperature haloaluminate ionic liquids: Novel solvents for transition metal solution chemistry. *Pure Appl. Chem.* 60: 1763–1772.
48. Paul, S. and Gupta, M. 2004. Selective Fries rearrangement catalyzed by zinc powder. *Synthesis* 1789–1792.
49. Miller, J. A. 1987. The metal-promoted Fries rearrangement. *J. Org. Chem.* 52: 322–323.
50. Baldwin, J. E. 1976. Rules for ring closure. *J. Chem. Soc., Chem. Commun.* 734–736.
51. Snieckus, V. 1990. Directed *ortho* metalation. Tertiary amide and *O*-carbamate directors in synthetic strategies for polysubstituted aromatics. *Chem. Rev.* 90: 879–933.
52. Cundy, C. S., Higgins, R., Kibby, S. A. M., Lowe, B. M., and Paton, R. M. 1989. *Para*-selective Fries rearrangement of phenyl acetate in the presence of zeolite molecular sieves. *Tetrahedron Lett.* 30: 2281–2284.
53. Vogt, A., Kouwenhoven, H. W., and Prins, R. 1995. Fries rearrangement over zeolitic catalysts. *Appl. Catal. A: Gen.* 123: 37–49.
54. Borzatta, V., Poluzzi, E., and Vaccari, A. 2002. In *Science and technology in catalysis*, ed. M. Anpo, M. Onaka, and H. Yamashita, 439–442, Amsterdam: Elsevier.
55. Borzatta, V., Busca, G., Poluzzi, E., Rossetti, V., Trombetta, M., and Vaccari, A. 2004. As to the reasons of the high activity of a commercial pentasil-type zeolite in the vapour-phase Fries rearrangement. *Appl. Catal. A: Gen.* 257: 85–95.
56. Armaroli, T., Bevilacqua, M., Trombetta, M., Gutièrrez Alejandre, A., Ramirez, J., and Busca, G. 2001. An FT-IR study of the adsorption of aromatic hydrocarbons and of 2,6-lutidine on H-FER and H-ZSM-5 zeolites. *Appl. Catal. A: Gen.* 220: 181–190.
57. Heidekum, A., Harmer, M. A., and Hoelderich, W. F. 1998. Highly selective Fries rearrangement over zeolites and Nafion in silica composite catalysts: a comparison. *J. Catal.* 176: 260–263.
58. Subba Rao, Y. V., Kulkarni, S. J., Subrahmanyam, M., and Rama Rao, A. V. 1993. Highly selective Fries rearrangement over modified ZSM-5 catalysts. *Tetrahedron Lett.* 34: 7799–7800.
59. Jayat, F., Sabater Picot, M. J., and Guisnet, M. 1996. Solvent effects in liquid phase Fries rearrangement of phenyl acetate over a HBea zeolite. *Catal. Lett.* 41: 181–187.
60. Fromentin, E., Coustard, J.-M., and Guisnet, M. 2000. Acetylation of 2-methoxynaphthalene with acetic anhydride over a HBEA zeolite. *J. Mol. Catal. A: Chem.* 159: 377–388.
61. Heitling, E., Roessner, F., and van Steen, E. 2004. Origin of catalyst deactivation in Fries rearrangement of phenyl acetate over zeolite H-Beta. *J. Mol. Catal. A: Chem.* 216: 61–65.
62. Freese, U., Heinrich, F., and Roessner, F. 1999. Acylation of aromatic compounds on H-Beta zeolites. *Catal. Today* 49: 237–244.

63. Wang, H. and Zou, Y. 2003. Modified Beta zeolite as catalyst for Fries rearrangement reaction. *Catal. Lett.* 86: 163–167.
64. Kozhevnikov, I. V. 2003. Friedel–Crafts acylation and related reactions catalysed by heteropoly acids. *Appl. Catal. A: Gen.* 256: 3–18.
65. Kozhevnikova, E. F., Derouane, E. G., and Kozhevnikov, I. V. 2002. Heteropoly acid as a novel efficient catalyst for Fries rearrangement. *Chem. Commun.* 1178–1179.
66. Kozhevnikova, E. F., Quartararo, J., and Kozhevnikov, I. V. 2003. Fries rearrangement of aryl esters catalysed by heteropoly acid. *Appl. Catal. A: Gen.* 245: 69–78.
67. Kozhevnikova, E. F., Rafiee, E., and Kozhevnikov, I. V. 2004. Fries rearrangement of aryl esters catalysed by heteropoly acid: catalyst regeneration and reuse. *Appl. Catal. A: Gen.* 260: 25–34.
68. Kaur, J., Kozhevnikova, E. F., Griffin, K., Harrison, B., and Kozhevnikov, I. V. 2003. Friedel–Crafts acylation and related reactions catalyzed by heteropoly acids. *Kinet. Catal.* 44: 175–182.
69. Kozhevnikov, I. V., Kloetstra, K. R., Sinnema, A., Zandbergen, H. W., and van Bekkum, H. 1996. Study of catalysts comprising heteropoly acid $H_3PW_{12}O_{40}$ supported on MCM-41 molecular sieve and amorphous silica. *J. Mol. Catal. A: Chem.* 114: 287–298.
70. Izumi, Y., Ono, M., Kitagawa, M., Yoshida, M., and Urabe K. 1995. Silica-included heteropoly compounds as solid acid catalysts. *Microporous Mater.* 5: 255–262.
71. Molnár, Á., Keresszegi, C., and Török, B. 1999. Heteropoly acids immobilized into a silica matrix: Characterization and catalytic applications. *Appl. Catal. A: Gen.* 189: 217–224.
72. Misono M. 2001. Unique acid catalysis of heteropoly compounds (heteropolyoxometalates) in the solid state. *Chem. Commun.* 1141–1152.
73. Venkatachalapathy, C. and Pitchumani, K. 1997. Fries rearrangement of esters in montmoriilonite clays: steric control on selectivity. *Tetrahedron* 33: 17171–17176.
74. Olah, G. A., Arvanaghi, M., and Krishnamurthy, V. V. 1983. Heterogeneous catalysis by solid superacids. 17. Polymeric perfluorinated resin sulfonic acid (Nafion-H) catalyzed Fries rearrangement of aryl esters. *J. Org. Chem.* 48: 3359–3360.
75. Pitchumani, K., Warrier, M., and Ramamurthy. V. 1996. Remarkable product selectivity during photo-Fries and photo-Claisen rearrangements within zeolites *J. Am. Chem. Soc.* 118: 9428–9429.
76. Hepp, M. A., Ramamurthy, V., Corbin, D. R., and Dybowski, C. 1992. Deuteron NMR investigations of ion–molecule interactions of aromatics included in zeolites. *J. Phys. Chem.* 96: 2629–2632.
77. Suau, R., Torres, G., and Valpuesta, M. 1995. The photo-Fries rearrangement of 2,5-disubstituted phenyl acetates. *Tetrahedron Lett.* 36: 1311–1314.
78. Pitchumani, K., Warrier, M., Cui, C., Weiss, R. G., and Ramamurthy, V. 1996. Photo-Fries reaction of naphthyl esters within zeolites. *Tetrahedron Lett.* 37: 6251–6254.
79. García, H., Primo, J., and Miranda, M. A. 1985. The photo-Fries rearrangement in the presence of potassium carbonate: a convenient synthesis of *ortho*-hydroxyacetophenones. *Synthesis* 901–902.

80. Gu, W., Warrier, M., Ramamurthy, V., and Weiss, R. G. 1999. Photo-Fries reactions of 1-naphthyl esters in cation-exchanged zeolite Y and polyethylene media. *J. Am. Chem. Soc.* 121: 9467–9468.
81. Singh, A. K. and Raghuraman, T. S. 1985. Photorearrangement of phenyl cinnamates under micellar environment. *Tetrahedron Lett.* 26: 4125–4128.
82. Belluš, D. and Hrdlovič, P. 1967. Photochemical rearrangement of aryl, vinyl, and substituted vinyl esters and amides of carboxylic acids. *Chem. Rev.* 67: 599–609.
83. Moghaddam, F. M., Ghaffarzadeh, M., and Abdi-Oskoui, S. H. 1999. Tandem Fries reaction-conjugate addition under microwave irradiation in dry media; one-pot synthesis of flavanones. *J. Chem. Res. (S)* 574–575.

chapter 6

Concluding remarks— future outlook

It is well recognized today that advances in preparative chemistry require the development of clean catalytic methodologies. Indeed, catalysts are the key for sustainable development in chemical industry. They allow synthesizing of products in a resource protective way, with less consumption of energy and with minimum formation of by-products and waste.[1]

One of the salient features of the original Friedel–Crafts acylation is the requirement of one or more equivalents of the Lewis acid catalyst that cannot be recovered and reused. In fact, strong Lewis acids such as aluminum chloride also complex to a very significant extent with the carbonyl group of the ketone product that consequently acts as a catalyst poison.[2] The workup requires hydrolysis of the final complex, leading to the loss of the catalyst and producing large amounts of corrosive waste streams.

The first significant improvement of synthetic Friedel–Crafts acylation is represented by the possibility to perform the reaction with catalytic amounts of Lewis acids such as iron trichloride or zinc chloride as well as metals and nonmetals (i.e., iron and iodine). Reactions require relatively high temperatures at which the hydrogen chloride evolution is rapid and dissociation of ketone–Lewis acid complex occurs easily, restoring the active catalyst.[3] Further progress toward the development of effective catalytic acylation processes is based on the application of metal triflates such as lanthanide triflates.[4] These Lewis acids were developed by Kobayashi as very promising catalysts to promote organic reactions, including Friedel–Crafts acylations. They are very active (1%–5% mol with respect to the acylating agent) and insensitive to the wetness of the reaction medium; consequently, they can be recovered and reused.[5]

The preceding results, however, describe examples of preparative methods in which the catalyst still represents the most expensive component of the reaction mixture. This creates economic and environmental problems that are particularly serious when transition metals and, in general, soluble metal derivatives are utilized as catalysts because the final product may adsorb some of the catalyst material. Presently, the upper tolerance limit for contamination of compounds destined for human consumption by metals and, in particular, transition metals is 5 ppm, and future regulation is expected to lower this threshold to the parts per billion (ppb) range. Quite similar problems are involved in the production of chemicals for material chemistry due to the high level of purity required for the final products.

There are two ways to prevent the releasing of catalytically active metals into commercial products:

1. If starting materials and products are volatile and stable enough, the reaction can be performed in a gas phase with a solid catalyst. But, unfortunately, the majority of fine chemicals and pharmaceuticals containing an aromatic ketone moiety is constituted of nonvolatile compounds that decompose when heated at high temperatures for long periods.
2. In this case, a heterogeneous catalyst can be used under batch conditions. This would allow removal of the catalyst by filtration or centrifugation, making possible the recovery and reuse of the catalyst and, consequently, an increase in catalyst lifetime.

To hit these targets, a wide variety of solid catalysts have been utilized in Friedel–Crafts acylation reactions. Quite promising synthetic results have been achieved, and the reaction can be efficiently conducted with heterogeneous catalysts. These studies have led to the industrialization of at least one process, showing that all the problems of reactivity, catalyst deactivation, and engineering can be solved by using zeolites as catalysts.[6]

Thus, the Rhodia company developed and applied an industrial process for the acylation of anisole with acetic anhydride over BEA zeolite to selectively give *para*-acetylanisole.[7] Furthermore, the reaction of isobutylbenzene with the same acylating agent leads with high yield and selectivity to *para*-acetyl isobutylbenzene, which is an intermediate in the synthesis of ibuprofen, an important anti-inflammatory drug.[8]

An additional significant advantage of heterogeneous catalysis is the possibility to utilize the catalyst with fixed-bed reactors in continuous-flow mode. Results of studies on this growing field of applied research suggest that solid catalysts, especially if they are applied in a structural way, play an important role in the so-called integrated approach to environmental protection and economical benefits. This approach includes, among others, integration of various process operations such as chemical reaction, separation, heat exchange, and product purification. The result of process integration is the reduction of investment costs, which is frequently combined with energy recovery and space saving. Such links are possible, beside others, with the use of special reactors such as the continuous-flow-packed bed or, better, monolithic reactors.

To this end, zeolites have been supported onto ceramic materials, giving macroscopic shapings utilizable in fixed-bed reactors.[9,10] These systems have been exploited in the acylation of anisole with carboxylic acids,[11,12] producing important results that merit continuous improvement.

However, it must be underlined that the most important industrial catalytic processes were developed by purely empirical methods and countless screening experiments. The complexity of the solid catalysts still represents a serious obstacle to the understanding of the structure–reactivity relationship.[13]

Vast improvement in the comprehension of the mode of catalyst operation in fundamental reactions, including Friedel–Crafts acylation, has been achieved more recently through in situ analyses. Indeed, by developing this approach, it is ensured that the active state of the catalyst can be characterized and that the data are relevant to the reaction mechanism. However, only a fraction of mechanistic analyses can be conducted under in situ conditions.

In some instances, there still exist conflicting reports about the surface acidity–catalytic activity correlation. These differences may arise not only from the use of different reaction conditions and different approaches to preparing or modifying the catalysts but also from a poor characterization of the materials employed. Indeed, the detailed physicochemical characterization of the catalytic materials, as well as the study of their interaction with reagents and products, still represents well-recognized problems in the use of heterogeneous catalysis for organic syntheses.

A further series of problems that require addressing in depth are the correct evaluation of solid catalyst efficiency and reusability,[14] as well as real heterogeneity and the possible contribution of leaching phenomena.[15]

In conclusion, although a great number of studies on the homogeneous and heterogeneous catalytic Friedel–Crafts acylations have been developed, a general solution of the efficient and environmentally acceptable large-scale production of aromatic ketones is still missing.

There is, however, space for future development, particularly based on the synergic cooperation between catalysis, physicochemical characterization, and engineering groups. Hopefully, along with this contribution, a multitude of new catalytic and ecoefficient Friedel–Crafts acylation reactions will be investigated, providing the instruments to develop new sustainable synthetic routes to aromatic ketones.

References

1. Sheldon, R. A. 1997. Catalysis and pollution prevention. *Chem. Ind.* 12–15.
2. Olah, G. A. 1964. *Friedel–Crafts and related reactions*. London: Interscience.
3. Pearson, D. E. and Buehler, C. A. 1972. Friedel–Crafts acylations with little or no catalyst. *Synthesis* 533–542.

4. Kawada, A., Mitamura, S., and Kobayashi, S. 1993. Lanthanide trifluoro-methanesulfonates as reusable catalysts: catalytic Friedel–Crafts acylation. *J. Chem. Soc., Chem. Commun.* 1157–1158.

5. Kobayashi, S. 1994. Rare earth metal trifluoromethanesulfonates as water-tolerant Lewis acid catalysts in organic synthesis. *Synlett* 689–701.

6. Sheldon, R. A. and Van Bekkum, H. 2001. *Fine chemical through heterogeneous catalysis.* Weinheim: Wiley-VCH.

7. Spagnol, M., Gilbert, L., Guillot, H., and Tirel, P. J. 1997. Acylation method for an aromatic compound. WO Patent 9,748,665.

8. Vogt, A. and Pfenninger, A. 1996. Process for the preparation of aromatic ketones. EP Patent 0,701,987.

9. Xu, M., Cheng, M., and Bao, X. 2000. Growth of ultrafine zeolite Y crystals on metakaolin microspheres. *Chem. Commun.* 1873–1874.

10. García-Martínez, J., Cazorla-Amorós, D., Linares-Solano, A., and Lin, Y. S. 2001. Synthesis and characterisation of MFI-type zeolites supported on carbon materials. *Microporous Mesoporous Mater.* 42: 255–268.

11. Beers, A. E. W., Nijhuis, T. A., Kapteijn, F., and Moulijn, J. A. 2001. Zeolite coated structures for the acylation of aromatics. *Microporous Mesoporous Mat.* 48: 279–284.

12. Nijhuis, T. A., Beers, A. E. W., Vergunst, T., Hoek, I., Kapteijn, F., and Moulijn, J. A. 2001. Preparation of monolithic catalysts. *Catal. Rev.* 43: 345–380.

13. Schlögl, R. 1993. Heterogeneous catalysis—still magic or already science? *Angew. Chem., Int. Ed. Engl.* 32: 381–383.

14. Gladysz, J. A. 2001. Recoverable catalysts: ultimate goals, criteria of evaluation, and the green chemistry interface. *Pure Appl. Chem.* 73: 1319–1324.

15. Lempers H. E. B. and Sheldon, R. A. 1998. The stability of chromium in CrAPO-5, CrAPO-11, and CrS-1 during liquid phase oxidations. *J. Catal.* 175: 62–69.

Index